McDougal Littell

# Geometry
## Concepts and Skills

**Larson   Boswell   Stiff**

**CHAPTER 6** # Resource Book

The Resource Book contains the wide variety of black-line masters available for Chapter 6. The blacklines are organized by lesson. Included are support materials for the teacher as well as practice, activities, applications, and assessment resources.

 **McDougal Littell**
A HOUGHTON MIFFLIN COMPANY
Evanston, Illinois • Boston • Dallas

## Contributing Authors

The authors wish to thank the following individuals for their contributions to the Chapter 6 Resource Book.

*Rebecca Salmon Glus*
*Patrick M. Kelly*
*Lynn Hisae Lafferty*
*Cheryl A. Leech*
*Michelle H. McCarney*
*Jessica Pflueger*
*Barbara L. Power*
*Joanne V. Ricci*

The authors wish to thank Meridian Creative Group for their contributions in creating this publication.

ISBN: 0-618-14044-1

456789-BHV-04

# Contents

# Contents

# Contents

## Description of Resources

This Chapter Resource Book is organized by lessons within the chapter in order to make your planning easier. The following materials are provided:

**Tips for New Teachers**  These teaching notes provide both new and experienced teachers with useful teaching tips for each lesson, including tips about common errors and inclusion.

**Parent Guide for Student Success**  This guide helps parents contribute to student success by providing an overview of the chapter along with questions and activities for parents and students to work on together.

**Strategies for Reading Mathematics**  The first page teaches reading strategies to be applied to the current chapter and to later chapters. The second page is a visual glossary of key vocabulary.

**Lesson Plans and Lesson Plans for Block Scheduling**  This planning template helps teachers select the materials they will use to teach each lesson from among the variety of materials available for the lesson. The block-scheduling version provides additional information about pacing.

**Warm-Up Exercises and Daily Homework Quiz**  The warm-ups cover prerequisite skills that help prepare students for a given lesson. The quiz assesses students on the content of the previous lesson. (Transparencies also available)

**Technology Activities and Keystrokes**  Keystrokes for the Geometry software and calculators are provided for each Technology Activity and Technology exercise in the Student Edition, along with additional Technology Activities to begin selected lessons.

**Practice A and B**  These exercises offer additional practice for the material in each lesson, including application problems. There are two levels of practice for each lesson: A (basic) and B (average).

**Reteaching with Practice**  These two pages provide additional instruction, worked-out examples, and practice exercises covering the key concepts and vocabulary in each lesson.

**Quick Catch-Up for Absent Students**  This handy form makes it easy for teachers to let students who have been absent know what to do for homework and which activities or examples were covered in class.

# Contents

**Learning Activities** These enrichment activities apply the math taught in the lesson in an interesting way that lends itself to group work.

**Real-Life Applications** Students apply the mathematics covered in each lesson to solve an interesting real-life problem.

**Quizzes** The quizzes can be used to assess student progress on two or three lessons.

**Brain Games Support** These blackline masters make it easier for students to record their work on selected activities in the Student Edition.

**Chapter Review Games and Activities** This worksheet offers fun practice at the end of the chapter and provides an alternative way to review the chapter content in preparation for the Chapter Tests.

**Chapter Tests A and B** These are tests that cover the most important skills taught in the chapter. There are two levels of test: A (basic) and B (average).

**SAT/ACT Chapter Test** This test also covers the most important skills taught in the chapter, but questions are in multiple-choice format. (See *Alternative Assessment* for multi-step problems.)

**Alternative Assessment with Rubrics and Math Journal** A journal exercise has students write about the mathematics in the chapter. A multi-step problem has students apply a variety of skills from the chapter and explain their reasoning. Solutions and a 4-point rubric are included.

**Project with Rubric** The project allows students to delve more deeply into a problem that applies the mathematics of the chapter. Teacher's notes and a 4-point rubric are included.

**Cumulative Review** These practice pages help students maintain skills from the current chapter and preceding chapters.

**Cumulative Test** This test covers the most important skills from the current chapter and preceding chapters. Cumulative Tests can be found in Chapters 3, 6, 9, and 11.

# *Tips for New Teachers*

For use with Chapter 6

## LESSON 6.1

**INCLUSION** To help your students understand the names of polygons, explain that the first part of the name gives the number of sides. Tell them that "tri" means three, "quad" means four, "pent" means five, and so on. Make sure that your students understand that an *n*-gon is what we call an *n*-sided polygon. Suggest that they draw a 9-sided, a 10-sided, and an 11-sided polygon in their notes and label them as a 9-gon, a 10-gon, and an 11-gon, respectively.

**TEACHING TIP** Once students know the properties of triangles, they can apply them to other figures that contain triangles. This is illustrated on page 305. The quadrilateral is divided into two triangles, leading to the conclusion that the sum of the measures of the angles in a quadrilateral is 360°. Students may use this approach to divide any polygon into non-overlapping triangles and determine the sum of the measures of the interior angles.

## LESSON 6.2

**TEACHING TIP** Demonstrate the properties illustrated in the theorems on pages 310–312, by yourself or with the aid of students, by using a large pair of identically labeled congruent parallelograms. Place them to show opposite sides congruent, opposite angles congruent, etc.

**COMMON ERROR** Since parallelograms have sides that slant (different from a rectangle or square), students may have trouble realizing that the diagonals are not congruent. Use the models mentioned above to show the diagonals have different lengths.

## LESSON 6.3

**TEACHING TIP** Before you present Theorem 6.8 to your students, you should quickly review the Same-Side Interior Angles Converse on page 138. This can lead into a nice discussion of the theorem. You could draw a quadrilateral and ask your students how they could apply the Same-Side Interior Angles Converse to show that a quadrilateral is a parallelogram.

**TEACHING TIP** Help students recognize that the theorems on pages 316–318 for proving a quadrilateral is a parallelogram are the converse of those given in the previous lesson.

**COMMON ERROR** When doing a problem similar to the Example on page 322, students may try to show only one pair of slopes equal. Caution them to make sure they are meeting the complete conditions stated in the theorem. Stress that students must show both pairs of opposite slopes are equal.

## LESSON 6.4

**TEACHING TIP** In Lesson 6.6, your students will learn that a square is a special type of rhombus and a special type of rectangle. However, you may want to stress these facts to your students in this lesson. Then you can point out that Theorems 6.10 and 6.11 also apply to a square since a square is a special type of rhombus and a special type of rectangle.

# Tips for New Teachers

**For use with Chapter 6**

## LESSON 6.5

**TEACHING TIP**  Together, all of the theorems in Lessons 6.2 through 6.5 can be confusing to students. Suggest that they write all of the theorems and any relevant figures on a separate piece of paper to refer to when they are reading through the text and working on the exercises.

**TEACHING TIP**  Lessons 6.2 through 6.4 dealt with properties of different types of parallelograms. To avoid confusion, stress to your students that by definition, a trapezoid is *not* a parallelogram. In general, the properties of parallelograms do not hold for trapezoids.

## LESSON 6.6

**INCLUSION**  The diagram at the top of page 337 is a diagram that students could easily sketch for themselves on scrap paper when taking a quiz or a test. Students could also draw a similar diagram using just the general shape of the quadrilateral. It can be an excellent aid, especially for those who have limited English proficiency.

**TEACHING TIP**  The summary charts in the exercises on page 339 can be very helpful. Consider having similar ones available for the class and let students pair up to complete them. Extra copies may be used as a summary review, a future quiz, or as an introduction to a lesson.

## Outside Resources

### BOOKS/PERIODICALS

Kennedy, Joe and Eric McDowell. "Geoboard Quadrilaterals." *Mathematics Teacher* (April 1998); pp. 288–290.

Housinger, Margret M. "Trap a Surprise in an Isosceles Trapezoid." *Mathematics Teacher* (January 1996); pp. 12–14.

### ACTIVITIES/MANIPULATIVES

Cuisenaire. *Geoboard Activity Mats.* White Plains, NY; Dale Seymour Publications.

### SOFTWARE

*Shape Makers: Developing Geometric Reasoning with the Geometer's Sketchpad.* Enables students to dynamically transform a shape to explore properties of triangles and quadrilaterals. Emeryville, CA; Key Curriculum Press.

NAME _____ DATE _____

# *Parent Guide for Student Success*

For use with Chapter 6

**Chapter Overview** One way that you can help your student to succeed in Chapter 6 is by discussing the lesson goals in the chart below. When a lesson is completed, you can ask your student the following questions. "What were the goals of the lesson? What new words and formulas did you learn? How can you apply the ideas of the lesson to your life?"

| *Lesson Title* | *Lesson Goals* | *Key Applications* |
|---|---|---|
| **6.1: Polygons** | Identify and classify polygons. Find angle measures of quadrilaterals. | • Parachutes<br>• Plants<br>• Envelopes |
| **6.2: Properties of Parallelograms** | Use properties of parallelograms. | • Scissors Lift<br>• Staircases |
| **6.3: Showing Quadrilaterals are Parallelograms** | Show that a quadrilateral is a parallelogram. | • Bicycle Gears<br>• Billiards |
| **6.4: Rhombuses, Rectangles, and Squares** | Use properties of special types of parallelograms. | • Carpentry<br>• Furniture Design |
| **6.5: Trapezoids** | Use properties of trapezoids. | • Cake Design |
| **6.6: Reasoning About Special Quadrilaterals** | Identify special quadrilaterals based on limited information. | • Gem Cutting |

## Visual Strategy

**Drawing Quadrilaterals** is the visual strategy featured in Chapter 6 (see page 302). Sometimes, to draw a special quadrilateral, it is easier to start with the diagonals. Draw two intersecting segments that have the properties of the diagonals for the type of quadrilateral you are drawing (diagonals bisect each other, diagonals are congruent, diagonals are perpendicular, etc.) Then use the endpoints of the segments as the vertices of your quadrilateral. The original segments are the diagonals of the quadrilateral.

NAME _____ DATE _____

# *Parent Guide for Student Success*

**For use with Chapter 6**

**Key Ideas** Your student can demonstrate understanding of key concepts by working through the following exercises with you.

| Lesson | Exercise |
|--------|----------|
| **6.1** | If you connect the tips of the arms of a starfish, what polygon is formed? |
| **6.2** | You are making a rope ladder like the one diagrammed on the right. To make sure the rungs are horizontal, *ABCD* must be a parallelogram. Which sides are congruent? Which angles are supplementary to ∠*DAB*? |
| **6.3** | After the rope ladder in the Exercise for 6.2 is made, how can you make sure *ABCD* is a parallelogram? What theorem could you use? |
| **6.4** | To make the rope ladder like the one diagrammed in the Exercise for Lesson 6.2 so that the rails are vertical and the rungs are horizontal, *ABCD* would need to be a rectangle. How could you guarantee that it is a rectangle? What theorem or corollary would you use? |
| **6.5** | A trapezoid has bases of lengths 4 and 14. What is the length of the midsegment of the trapezoid? |
| **6.6** | The diagonals of quadrilateral *JKLM* are congruent. What quadrilaterals always meet this condition? What quadrilaterals sometimes meet this condition? |

## Home Involvement Activity

**You Will Need:** Materials to make sketches

**Directions:** Challenge your student to a quadrilateral scavenger hunt. Try to find and sketch a real-life example of each of the following: a square, a rectangle that is not a square, a rhombus that is not a square, a parallelogram that is not a rhombus or a rectangle, a trapezoid, and a quadrilateral that is none of these. See who can finish first.

**Answers**

**6.1:** pentagon **6.2:** $\overline{AB} \cong \overline{DC}$, $\overline{AD} \cong \overline{BC}$, ∠*ADC* and ∠*ABC* **6.3:** *Sample answer:* make sure $AB = DC$ and $AD = BC$; Theorem 6.6 (See p. 319.) **6.4:** make sure $AC = BD$; Theorem 6.11 (See p. 327.) **6.5:** 9 **6.6:** always: rectangle, square; sometimes: parallelogram, rhombus

NAME _____ DATE _____

# *Strategies for Reading Mathematics*

**For use with Chapter 6**

## *Strategy: Reading Symbols*

Symbols are an important part of geometry. They help you communicate ideas
without using words. You need to know the meaning of the symbols in order to
use and understand the ideas being communicated. Both sets of statements
below are the same, but the ones on the left use words and the ones on the right
use symbols.

| *Words* | *Symbols* |
|---|---|
| The measure of angle *A* is equal to the measure of angle *D*. | $m\angle A = m\angle D$ |
| Angle *A* is congruent to angle *D*. | $\angle A \cong \angle D$ |
| The length of segment *AB* is equal to the length of segment *DE*. | $AB = DE$ |
| Segment *AB* is congruent to segment *DE*. | $\overline{AB} \cong \overline{DE}$ |
| Triangle *ABC* is congruent to triangle *DEF*. | $\triangle ABC \cong \triangle DEF$ |

> **STUDY TIP**
> *Using Symbols*
> Pay special attention to the order of the
> vertices in statements about congruent
> figures. Look at the vertices listed first.
> They are vertices of two congruent angles.
> Find the other pairs of congruent angles
> by looking at the vertices listed second,
> the vertices listed third, and so on.

## Questions

1. Look at the first and second pairs of statements given above. What is
   the difference between $\angle A$ and $m\angle A$? When would you use the symbol
   $\angle A$? When would you use the symbol $m\angle A$?

2. Look at the third and fourth pairs of statements given above. What is
   the difference between $\overline{AB}$ and $AB$? When would you use the symbol
   $\overline{AB}$? When would you use the symbol $AB$?

3. What is the difference between the symbols = and $\cong$? When would
   you use the symbol = in a geometric statement? When would you use
   the symbol $\cong$?

4. Is the statement $\triangle ABC \cong \triangle DEF$ the same as the statement
   $\triangle ABC \cong \triangle EFD$? Explain.

NAME_____ DATE _____

# *Strategies for Reading Mathematics*

**For use with Chapter 6**

## *Visual Glossary*

The Study Guide on page 302 lists the key words for Chapter 6. Use the visual glossary below to help you understand some of the key words in Chapter 6. You may want to copy these diagrams into your notebook and refer to them as you complete the chapter.

### GLOSSARY

**polygon** (p. 303) A plane figure that is formed by three or more segments called *sides*. Each side intersects exactly two other sides, one at each of its endpoints. Each endpoint is a *vertex* of the polygon.

**quadrilateral** (p. 304) A polygon with four sides.

**parallelogram** (p. 310) A quadrilateral with both pairs of opposite sides parallel. The symbol for parallelogram *ABCD* is ▱*ABCD*.

**rhombus** (p. 325) A parallelogram with four congruent sides.

**rectangle** (p. 325) A parallelogram with four right angles.

**square** (p. 325) A parallelogram with four congruent sides and four right angles.

**trapezoid** (p. 332) A quadrilateral with exactly one pair of parallel sides, called *bases*. The nonparallel sides are the *legs*.

### Interior Angles of Polygons

The sum of the interior angles of a triangle is 180°. This information can be used to find the sum of the interior angles of polygons with more than three sides by dividing up the polygon into non-overlapping triangles.

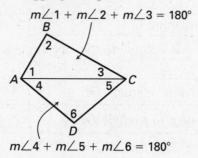

Sum of interior angles of quadrilateral $ABCD = 180° + 180° = 360°$.

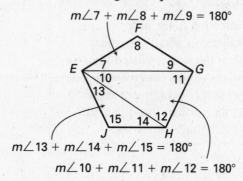

Sum of interior angles of pentagon $EFGHJ = 180° + 180° + 180° = 540°$.

### Special Quadrilaterals

Quadrilaterals with certain properties are given special names. Some special quadrilaterals are shown below.

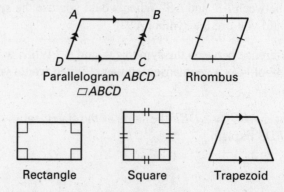

Parallelogram *ABCD*
▱*ABCD*

Rhombus

Rectangle    Square    Trapezoid

**LESSON 6.1**

# Lesson Plan

2-day lesson (See *Pacing the Chapter*, TE page 300A)                    For use with pages 303–308

**GOAL** Identify and classify polygons. Find angle measures of quadrilaterals.

State/Local Objectives _____

_____

✓ **Check the items you wish to use for this lesson.**

**STARTING OPTIONS**

____ Strategies for Reading Mathematics: CRB pages 5–6

____ Warm-Up: CRB page 9 or Transparencies

**TEACHING OPTIONS**

____ Examples: Day 1: 1–2, SE pages 303–304; Day 2: 3, SE page 305

____ Extra Examples: TE pages 304–305

____ Checkpoint Exercises: Day 1: 1–4, SE page 304; Day 2: 5–7, SE page 305

____ Technology Keystrokes for Ex. 28 on SE page 308: CRB page 10

____ Concept Check: TE page 305

____ Guided Practice Exercises: Day 1: 1–5, SE page 306; Day 2: 6–7, SE page 306

**APPLY/HOMEWORK**

**Homework Assignment**

____ Basic: Day 1: pp. 306–308 Exs. 8–14, 21, 22, 30–42 even

Day 2: pp. 306–308 Exs. 15–20, 29–41 odd

____ Average: Day 1: pp. 306–308 Exs. 8–14, 21–27, 30–36 even

Day 2: pp. 306–308 Exs. 15–20, 28, 29–35 odd, 37–42

____ Advanced: Day 1: pp. 306–308 Exs. 8–14, 21–27, 30–42 even

Day 2: pp. 306–308 Exs. 15–20, 28, 29–41 odd; EC: TE p. 300D*, classzone.com

**Reteaching the Lesson**

____ Practice Masters: CRB pages 11–12 (Level A, Level B)

____ Reteaching with Practice: CRB pages 13–14 or Practice Workbook with Examples;

Resources in Spanish

**Extending the Lesson**

____ Challenge: TE page 300D; classzone.com

**ASSESSMENT OPTIONS**

____ Daily Quiz (6.1): TE page 308, CRB page 18, or Transparencies

____ Standardized Test Practice: SE page 308; Transparencies

Notes _____

_____

_____

TEACHER'S NAME _____ CLASS _____ ROOM _____ DATE _____

# Lesson Plan for Block Scheduling

1-block lesson (See *Pacing the Chapter,* TE page 300A)          For use with pages 303–308

**GOAL**   Identify and classify polygons. Find angle measures of quadrilaterals.

State/Local Objectives _____

_____

_____

| CHAPTER PACING GUIDE | |
|---|---|
| **Day** | **Lesson** |
| 1 | **6.1** |
| 2 | 6.2 |
| 3 | 6.3 |
| 4 | 6.4 |
| 5 | 6.5 |
| 6 | 6.6 |
| 7 | Ch. 6 Review and Assess |

✓ **Check the items you wish to use for this lesson.**

**STARTING OPTIONS**
_____ Strategies for Reading Mathematics: CRB pages 5–6
_____ Warm-Up: CRB page 9

**TEACHING OPTIONS**
_____ Examples: 1–3, SE pages 303–305
_____ Extra Examples: TE pages 304–305
_____ Checkpoint Exercises: 1–7, SE pages 304–305
_____ Technology Keystrokes for Ex. 28 on page 308: CRB page 10
_____ Concept Check: TE page 305
_____ Guided Practice Exercises: 1–7, SE page 306

**APPLY/HOMEWORK**
**Homework Assignment**
_____ Block Schedule: pp. 306–308 Exs. 8–42

**Reteaching the Lesson**
_____ Practice Masters: CRB pages 11–12 (Level A, Level B)
_____ Reteaching with Practice: CRB pages 13–14 or Practice Workbook with Examples;
Resources in Spanish

**Extending the Lesson**
_____ Challenge: TE page 300D; classzone.com

**ASSESSMENT OPTIONS**
_____ Daily Quiz (6.1): TE page 308, CRB page 18, or Transparencies
_____ Standardized Test Practice: SE page 308; Transparencies

Notes _____

_____

_____

**Geometry**
Chapter 6 Resource Book

Lesson 6.1

NAME _____ DATE _____

# WARM-UP EXERCISES

For use before Lesson 6.1, pages 303–308

1. What is the sum of the measures of the interior angles of a triangle?

2. In which kind of triangle are all three sides congruent?

3. In which kind of triangle are all three angles congruent?

## DAILY HOMEWORK QUIZ

For use after Lesson 5.7, pages 281–290

**Determine whether figure B is a reflection of figure A.**

1.

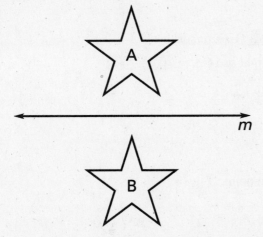

2.

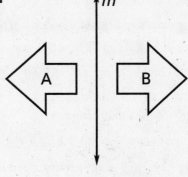

3. Determine whether the grid shows a reflection in the *x-axis*, the *y-axis*, or *neither*.

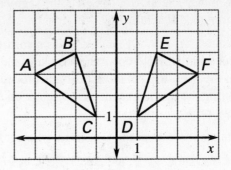

4. Find the number of lines of symmetry.

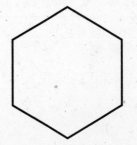

## Keystrokes for Exercise 28

### TI-92

1. Draw quadrilateral *ABCD*.

   **F3** 4 (Place cursor at desired location.) **ENTER** *A* (Place cursor at desired location.) **ENTER** *B* (Place cursor at desired location.) **ENTER** *C* (Place cursor at desired location.) **ENTER** *D* (Move cursor to point *A*.) **ENTER**

2. Measure ∠*A*, ∠*B*, ∠*C*, and ∠*D*.

   **F6** 3 (Move cursor to point *D*.) **ENTER** (Move cursor to point *A*.) **ENTER** (Move cursor to point *B*.) **ENTER** Repeat this procedure for the other angles.

3. Calculate the sum of the four angles.

   **F6** 6 (Use cursor to highlight *m*∠*A*.) **ENTER** **+** (Use cursor to highlight *m*∠*B*.) **ENTER** **+** (Use cursor to highlight *m*∠*C*.) **ENTER** **+** (Use cursor to highlight *m*∠*D*.) **ENTER** **ENTER** (The result will appear on the screen.)

4. Drag a vertex of the quadrilateral.

   **F1** 1 (Move cursor to point *A*.) **ENTER**

   Use the drag key [⟲] and the cursor pad to drag the point.

### SKETCHPAD

1. Draw quadrilateral *ABCD*. Select the segment tool. Draw $\overline{AB}$, $\overline{BC}$, $\overline{CD}$, and $\overline{DA}$ to form quadrilateral *ABCD*.

2. Measure ∠*A*, ∠*B*, ∠*C*, and ∠*D*. To measure ∠*A*, choose the selection arrow tool. Select point *D*. Hold down the shift key, and select points *A* and *B*. Choose **Angle** from the **Measure** menu. Repeat this procedure for the remaining angles. Before selecting the next angle, be sure to click on a blank area of the window to deselect the previous points.

3. Calculate the sum of the four angles. Select **Calculate** from the **Measure** menu. Click *m*∠*DAB*, **+** , *m*∠*ABC*, **+** , *m*∠*BCD*, **+** , *m*∠*CDA*, and **OK** .

4. Choose the selection arrow tool. Select and drag a vertex of the quadrilateral.

NAME _____ DATE _____

# Practice A

For use with pages 303–308

**Match the key word with its description.**

1. sides of a polygon

2. diagonal of a polygon

3. polygon

4. vertex of a polygon

A. a plane figure that is formed by three or more segments

B. the segments that form a polygon

C. each endpoint of a side of a polygon

D. a segment that joins two nonconsecutive vertices of a polygon

**Is the figure a polygon? Explain your reasoning.**

5.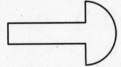

6.

7.

**Classify the polygon by its number of sides.**

8.

9.

10.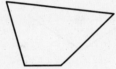

**Find the measure of ∠K.**

11.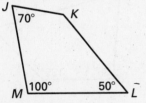

12.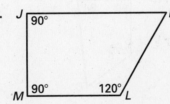

13.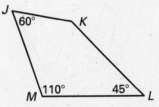

**Find the value of x.**

14.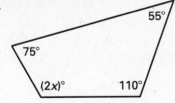

15.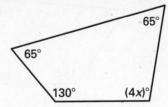

16.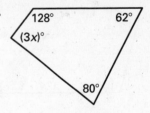

**The game board pictured at the right has the shape of a polygon.**

17. Tell how many sides the polygon has and what type of polygon it is.

18. Name the vertices of the polygon.

19. Name the sides of the polygon.

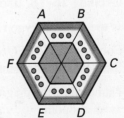

## Practice B

For use with pages 303–308

**Polygon ABCDE is shown at the right.**

1. Name the vertices of polygon *ABCDE*.

2. Name the sides of polygon *ABCDE*.

3. Name the diagonals from vertex *D* of polygon *ABCDE*.

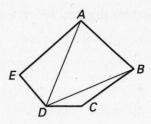

**Decide whether the figure is a polygon. If so, tell what type. If not, explain why.**

4.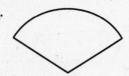

5.

6.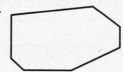

7.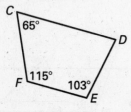

8.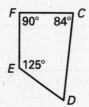

9.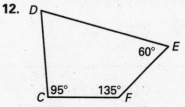

**Find the measure of ∠D.**

10.

11.

12.

**Find the value of the variable.**

13.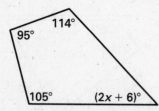

14.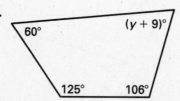

15.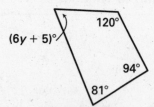

**In Exercises 16–18, use the stop sign pictured at the right.**

16. Is the sign a polygon? If so, tell what type. If not, explain why.

17. Polygon *ABCDEFGH* is one name for the polygon. State two other names using the vertices.

18. Name all of the diagonals of the polygon that have vertex *G* as an endpoint.

NAME_____ DATE _____

# Reteaching with Practice

For use with pages 303–308

**GOAL** Identify and classify polygons. Find angle measures of quadrilaterals.

## VOCABULARY

A **polygon** is a plane figure that is formed by three or more segments called **sides**. The endpoint of each side is a **vertex**.

A segment that joins two nonconsecutive vertices of a polygon is called a **diagonal**.

Polygons are classified by the number of sides they have. A **triangle** has three sides. A **quadrilateral** has four sides. A **pentagon** has five sides. A **hexagon** has six sides. A **heptagon** has seven sides. An **octagon** has eight sides.

**Theorem 6.1   Quadrilateral Interior Angles Theorem**
The sum of the measures of the interior angles of a quadrilateral is 360°.

**EXAMPLE 1** *Identify Polygons*

Tell whether the figure is a polygon. Explain your reasoning.

a.    b. 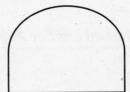   c.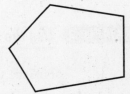

## SOLUTION

a. No, the figure is not a polygon because each side intersects two other sides at one vertex, and no other sides at the other vertex.

b. No, the figure is not a polygon because it has a side that is not a segment.

c. Yes, the figure is a polygon formed by five straight sides.

### Exercises for Example 1

**Tell whether the figure is a polygon. Explain your reasoning.**

1. 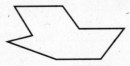   2.    3.

4.    5.

# *Reteaching with Practice*

For use with pages 303–308

**EXAMPLE 2** *Classify Polygons*

Decide whether the figure is a polygon. If so, tell what type. If not, explain why.

a. 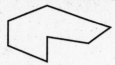  b. 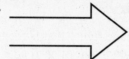  c.

### SOLUTION

**a.** The figure is a polygon with seven sides, so it is a heptagon.

**b.** The figure is not a polygon because two of the sides intersect only one other side.

**c.** The figure is a polygon with six sides, so it is a hexagon.

### *Exercises for Example 2*

**Decide whether the figure is a polygon. If so, tell what type. If not, explain why.**

6.    7.    8.

**EXAMPLE 3** *Use the Quadrilateral Interior Angles Theorem*

Find the value of $x$.

### SOLUTION

Use the fact that the sum of the measures of the interior
angles of a quadrilateral is 360°.

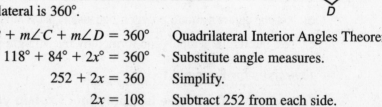

$m\angle A + m\angle B + m\angle C + m\angle D = 360°$ — Quadrilateral Interior Angles Theorem

$50° + 118° + 84° + 2x° = 360°$ — Substitute angle measures.

$252 + 2x = 360$ — Simplify.

$2x = 108$ — Subtract 252 from each side.

$x = 54$ — Divide each side by 2.

### *Exercises for Example 3*

**Find the value of $x$.**

9. 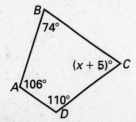   10. 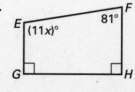   11.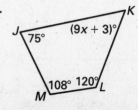

NAME_____ DATE _____

# *Quick Catch-Up for Absent Students*

**For use with pages 303–308**

The items checked below were covered in class on (date missed) _____

**Lesson 6.1: Polygons  (pp. 303–305)**

____ **Goal:** Identify and classify polygons. Use the Quadrilateral Interior Angles Theorem.

*Material Covered:*

____ Student Help: Vocabulary Tip

____ Example 1: Identify Polygons

____ Student Help: Study Tip

____ Example 2: Classify Polygons

____ Student Help: Study Tip

____ Example 3: Find Angle Measures of Quadrilaterals

*Vocabulary:*

polygon, p. 303                           vertex of a polygon, p. 303

side of a polygon, p. 303                 diagonal of a polygon, p. 303

____ Other (specify) _____

_____

**Homework and Additional Learning Support**

____ Textbook exercises (teacher to specify) pp. 306–308 _____

_____

____ Internet: Homework Help at classzone.com

____ *Reteaching with Practice* worksheet

TEACHER'S NAME _____ CLASS _____ ROOM _____ DATE _____

# *Lesson Plan*

**2-day lesson (See *Pacing the Chapter*, TE page 300A)**

**GOAL** Use properties of parallelograms.

State/Local Objectives _____

_____

## ✓ Check the items you wish to use for this lesson.

### STARTING OPTIONS
_____ Homework Check (6.1): TE page 306; Answer Transparencies
_____ Homework Quiz (6.1): TE page 308, CRB page 18, or Transparencies
_____ Warm-Up: CRB page 18 or Transparencies

### TEACHING OPTIONS
_____ Activity: SE page 309
_____ Examples: Day 1: 1–2, SE pages 310–311; Day 2: 3, SE page 312
_____ Extra Examples: TE pages 311–312
_____ Checkpoint Exercises: Day 1: 1–3, SE pages 310–311
_____ Concept Check: TE page 312
_____ Guided Practice Exercises: Day 1: 1–6, SE page 313; Day 2: 7–12, SE page 313
_____ Visualize It! Transparencies: 26, 27

### APPLY/HOMEWORK
**Homework Assignment**
_____ Basic: Day 1: pp. 313–315 Exs. 14, 16–20, 22–24, 31, 32, 43, 44–54 even
　　　　Day 2: pp. 313–315 Exs. 13, 15, 25–30, 42, 45–55 odd
_____ Average: Day 1: pp. 313–315 Exs. 14, 16–24, 31, 32, 38–40, 43, 44–54 even
　　　　Day 2: pp. 313–315 Exs. 13, 15, 25–30, 33–37, 42, 45–55 odd
_____ Advanced: Day 1: pp. 313–315 Exs. 14, 16–24, 31, 32, 38–40, 41*, 43, 44, 46
　　　　Day 2: pp. 313–315 Exs. 13, 15, 25–30, 33–37, 42, 47–55 odd; EC: classzone.com

**Reteaching the Lesson**
_____ Practice Masters: CRB pages 19–20 (Level A, Level B)
_____ Reteaching with Practice: CRB pages 21–22 or Practice Workbook with Examples;
　　　　Resources in Spanish

**Extending the Lesson**
_____ Challenge: SE page 315; classzone.com

### ASSESSMENT OPTIONS
_____ Daily Quiz (6.2): TE page 315, CRB page 26, or Transparencies
_____ Standardized Test Practice: SE page 315; Transparencies

Notes _____

_____

TEACHER'S NAME _____ CLASS _____ ROOM _____ DATE _____

# Lesson Plan for Block Scheduling

**1-block lesson (See *Pacing the Chapter*, TE page 300A)**          **For use with pages 309–315**

**GOAL**     **Use properties of parallelograms.**

State/Local Objectives _____

_____

_____

✓ **Check the items you wish to use for this lesson.**

**STARTING OPTIONS**

____ Homework Check (6.1): TE page 306; Answer Transparencies
____ Homework Quiz (6.1): TE page 308, CRB page 18,
       or Transparencies
____ Warm-Up: CRB page 18 or Transparencies

| CHAPTER PACING GUIDE | |
|---|---|
| **Day** | **Lesson** |
| 1 | 6.1 |
| 2 | **6.2** |
| 3 | 6.3 |
| 4 | 6.4 |
| 5 | 6.5 |
| 6 | 6.6 |
| 7 | Ch. 6 Review and Assess |

**TEACHING OPTIONS**

____ Activity: SE page 309
____ Examples: 1–3, SE pages 310–312
____ Extra Examples: TE pages 311–312
____ Checkpoint Exercises: 1–3, SE pages 310–311
____ Concept Check: TE page 312
____ Guided Practice Exercises: 1–12, SE page 313
____ Visualize It! Transparencies: 26, 27

**APPLY/HOMEWORK**
**Homework Assignment**

____ Block Schedule:  pp. 313–315 Exs. 13–40, 42–55

**Reteaching the Lesson**

____ Practice Masters: CRB pages 19–20 (Level A, Level B)
____ Reteaching with Practice: CRB pages 21–22 or Practice Workbook with Examples;
       Resources in Spanish

**Extending the Lesson**

____ Challenge: SE page 315; classzone.com

**ASSESSMENT OPTIONS**

____ Daily Quiz (6.2): TE page 315, CRB page 26, or Transparencies
____ Standardized Test Practice: SE page 315; Transparencies

Notes _____

_____

_____

_____

NAME _____ DATE _____

# WARM-UP EXERCISES

For use before Lesson 6.2, pages 309–315

**1.** Given that $\angle P$ is a supplement of $\angle R$ and $m\angle P = 164°$, find $m\angle R$.

**2.** Find the value of $x$.

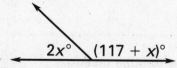

$2x°$ $(117 + x)°$

.........................................................................

# DAILY HOMEWORK QUIZ

For use after Lesson 6.1, pages 303–308

**Decide whether the figure is a polygon. If so, tell what type. If not, explain why.**

**1.**

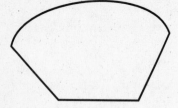

**2.**

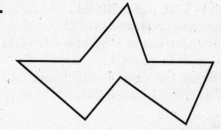

**Find the measure of ∠A.**

**3.**

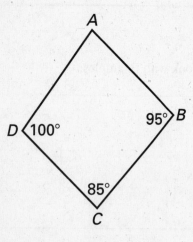

$A$

$D$ $100°$

$95°$ $B$

$85°$

$C$

**4.**

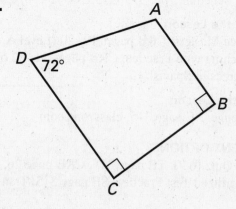

$A$

$D$ $72°$

$B$

$C$

**5.** Find the value of $x$.

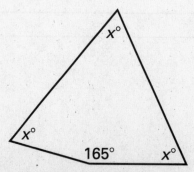

$x°$

$x°$

$165°$ $x°$

**Geometry**
Chapter 6 Resource Book

**Decide whether the figure is a parallelogram. If it is not, explain why.**

1.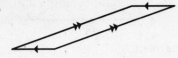

2.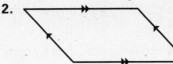

**JKLM is a parallelogram. Find JK and KL.**

3.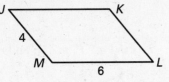

4.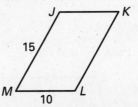

5.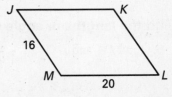

**WXYZ is a parallelogram. Find the missing angle measures.**

6.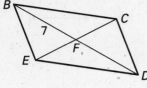

7.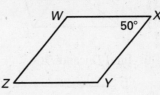

8.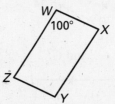

**BCDE is a parallelogram. Find DF.**

9.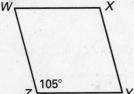

10.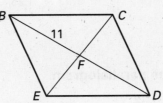

11.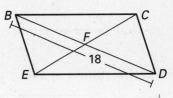

**Find the values of x and y in the parallelogram.**

12.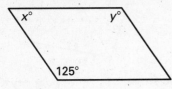

13.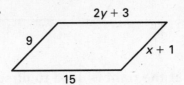

14.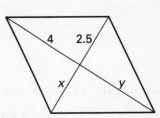

**The diagram at the right shows an adjustable coat rack.**

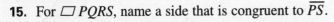
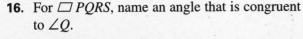

15. For ▱ *PQRS*, name a side that is congruent to $\overline{PS}$.

16. For ▱ *PQRS*, name an angle that is congruent to ∠*Q*.

17. The points *Q*, *U*, *T*, and *S* form a parallelogram. Use what you know about the diagonals of a parallelogram to name a length equal to *RU*.

Lesson 6.2

NAME_____ DATE _____

## Practice B

For use with pages 309–315

Write the statement of each theorem in symbols for ▱*PQRS,*
where *m∠SPQ* = *m∠QRS* = *x*° and *m∠RSP* = *m∠PQR* = *y*°.

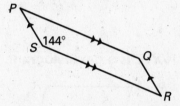

1. If a quadrilateral is a parallelogram, then its opposite sides are congruent.

2. If a quadrilateral is a parallelogram, then its opposite angles are congruent.

3. If a quadrilateral is a parallelogram, then its consecutive angles are supplementary.

4. If a quadrilateral is a parallelogram, then its diagonals bisect each other.

**Find the lengths or angle measures.**

5. Find *DA* and *DC.*

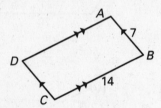

6. Find *GE* and *DF.*

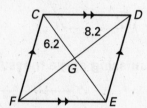

7. Find *m∠P, m∠Q,* and *m∠R.*

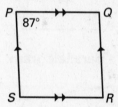

8. Find *GE* and *GF.*

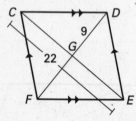

9. Find *DA* and *DC.*

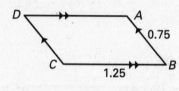

10. Find *m∠Q, m∠R,* and *m∠S.*

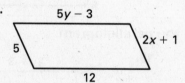

**Find the values of *x* and *y* in the parallelogram.**

11.

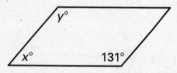

12.

13.

**The chevron symbol shown at the right is used to direct traffic flow.**

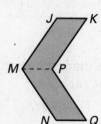

14. For ▱ *JKPM,* name two pairs of congruent sides.

15. For ▱ *MPQN,* name two pairs of congruent angles.

16. For ▱ *JKPM,* name two angles that are supplementary to ∠*K.*

Lesson 6.2

NAME_____ DATE _____

# *Reteaching with Practice*

**For use with pages 309–315**

**GOAL** **Use properties of parallelograms.**

---

**VOCABULARY**

A **parallelogram** is a quadrilateral with both pairs of opposite sides parallel.

**Theorem 6.2**
If a quadrilateral is a parallelogram, then its opposite sides are congruent.

**Theorem 6.3**
If a quadrilateral is a parallelogram, then its opposite angles are congruent.

**Theorem 6.4**
If a quadrilateral is a parallelogram, then its consecutive angles are supplementary.

**Theorem 6.5**
If a quadrilateral is a parallelogram, then its diagonals bisect each other.

---

**EXAMPLE 1** *Find Side Lengths of Parallelograms*

*ABCD* is a parallelogram.

Find the values of *x* and *y*.

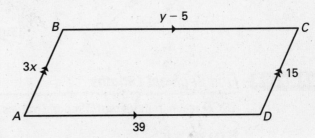

**SOLUTION**

| | |
|---|---|
| $AB = CD$ | Opposite sides of a parallelogram are congruent. |
| $3x = 15$ | Substitute $3x$ for $AB$ and 15 for $CD$. |
| $x = 5$ | Divide each side by 3. |
| $BC = AD$ | Opposite sides of a parallelogram are congruent. |
| $y - 5 = 39$ | Substitute $y - 5$ for $BC$ and 39 for $AD$. |
| $y = 44$ | Add 5 to each side. |

**Exercises for Example 1**

.....................................................................................................................

**Find the values of *x* and *y* in the parallelogram.**

**1.**

**2.**
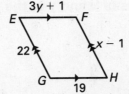

**3.**
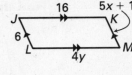

Lesson 6.2

NAME _____  DATE _____

# *Reteaching with Practice*

For use with pages 309–315

**EXAMPLE 2** *Find Angle Measures of Parallelograms*

*ABDC* is a parallelogram. Find the values of *x* and *y*.

**SOLUTION**

By Theorem 6.4, the consecutive angles of a parallelogram are supplementary.

| | |
|---|---|
| $m\angle A + m\angle C = 180°$ | Theorem 6.4 |
| $72° + x° = 180°$ | Substitute 72° for $m\angle A$ and $x°$ for $m\angle C$. |
| $x = 108$ | Subtract 72 from each side. |

By Theorem 6.3, the opposite angles of a parallelogram are congruent.

| | |
|---|---|
| $m\angle A = m\angle D$ | Opposite angles of a □ are congruent. |
| $72° = 3y°$ | Substitute 72° for $m\angle A$ and $3y°$ for $m\angle D$. |
| $24 = y$ | Divide each side by 3. |

## Exercises for Example 2

**Find the values of *x* and *y* in the parallelogram.**

**4.**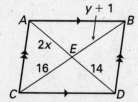

**5.**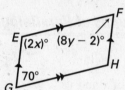

**6.**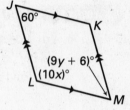

**EXAMPLE 3** *Find Segment Lengths*

*ABCD* is a parallelogram. Find the values of *x* and *y*.

**SOLUTION**

By Theorem 6.5, the diagonals bisect each other.

| | |
|---|---|
| $BE = DE$ | Diagonals of a □ bisect each other. |
| $6 = x$ | Substitute 6 for *BE* and *x* for *DE*. |

Use Theorem 6.5 again for the other diagonal.

| | |
|---|---|
| $AE = CE$ | Diagonals of a □ bisect each other. |
| $9 = 3y$ | Substitute 9 for *AE* and 3y for *CE*. |
| $3 = y$ | Divide each side by 3. |

## Exercises for Example 3

**Find the values of *x* and *y* in the parallelogram.**

**7.**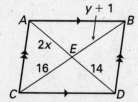

**8.**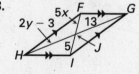

**9.**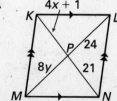

Lesson 6.2

NAME _____ DATE _____

# Quick Catch-Up for Absent Students

**For use with pages 309–315**

The items checked below were covered in class on (date missed) _____

**Activity 6.2: Investigating Parallelograms (p. 309)**

____ **Goal:** Find some of the properties of a parallelogram.

**Lesson 6.2: Properties of Parallelograms (pp. 310–312)**

____ **Goal:** Use properties of parallelograms.

*Material Covered:*

____ Example 1: Find Side Lengths of Parallelograms

____ Example 2: Find Angle Measures of Parallelograms

____ Student Help: Look Back

____ Example 3: Find Segment Lengths

*Vocabulary:*

parallelogram, p. 310

____ Other (specify) _____

_____

**Homework and Additional Learning Support**

____ Textbook exercises (teacher to specify) pp. 313–315 _____

_____

____ Internet: Homework Help at classzone.com

____ *Reteaching with Practice* worksheet

TEACHER'S NAME _____ CLASS _____ ROOM _____ DATE _____

## *Lesson Plan*

2-day lesson (See *Pacing the Chapter,* TE page 300A)

For use with pages 316–324

**GOAL** **Show that a quadrilateral is a parallelogram.**

State/Local Objectives _____

_____

✓ **Check the items you wish to use for this lesson.**

**STARTING OPTIONS**
_____ Homework Check (6.2): TE page 313; Answer Transparencies
_____ Homework Quiz (6.2): TE page 315, CRB page 26, or Transparencies
_____ Warm-Up: CRB page 26 or Transparencies

**TEACHING OPTIONS**
_____ Geo-Activity: SE page 316
_____ Examples: Day 1: 1–2, SE page 317; Day 2: 3–4, SE page 318
_____ Extra Examples: TE pages 317–318; Internet Help at classzone.com
_____ Checkpoint Exercises: Day 1: 1–3, SE page 317; Day 2: 4–7, SE page 319
_____ Technology Activity: SE page 324
_____ Technology Keystrokes for Activity on SE page 324: CRB pages 27–28
_____ Concept Check: TE page 319
_____ Guided Practice Exercises: Day 1: 1–2, 6–7, SE page 320; Day 2: 3–5, SE page 320
_____ Visualize It! Transparencies: 28

**APPLY/HOMEWORK**
**Homework Assignment**
_____ Basic: Day 1: pp. 320–323 Exs. 8–13, 21, 27–40
      EP: pp. 679–680 Exs. 21, 28, 31, 34; pp. 320–323 Exs. 14–19, 25, 26, Quiz 1
_____ Average: Day 1: pp. 320–323 Exs. 8–13, 20, 21, 25–38
      Day 2: pp. 320–323 Exs. 14–19, 22, 23, 39–44, Quiz 1
_____ Advanced: Day 1: pp. 320–323 Exs. 8–13, 20, 21, 25–34
      Day 2: pp. 320–323 Exs. 14–19, 22, 23, 24*, 35–43 odd, Quiz 1; EC: classzone.com

**Reteaching the Lesson**
_____ Practice Masters: CRB pages 29–30 (Level A, Level B)
_____ Reteaching with Practice: CRB pages 31–32 or Practice Workbook with Examples;
      Resources in Spanish

**Extending the Lesson**
_____ Challenge: SE page 321; classzone.com

**ASSESSMENT OPTIONS**
_____ Daily Quiz (6.3): TE page 323, CRB page 37, or Transparencies
_____ Standardized Test Practice: SE page 322; Transparencies
_____ Quiz (6.1–6.3): SE page 323; CRB page 34

Notes _____

_____

_____

**Geometry**
Chapter 6 Resource Book

TEACHER'S NAME _____ CLASS _____ ROOM _____ DATE _____

# *Lesson Plan for Block Scheduling*

1-block lesson (See *Pacing the Chapter,* TE page 300A)          **For use with pages 316–324**

**GOAL**   **Show that a quadrilateral is a parallelogram.**

State/Local Objectives _____

_____

_____

✓ **Check the items you wish to use for this lesson.**

## STARTING OPTIONS

____ Homework Check (6.2): TE page 313; Answer Transparencies
____ Homework Quiz (6.2): TE page 315, CRB page 26, or Transparencies
____ Warm-Up: CRB page 26 or Transparencies

## TEACHING OPTIONS

____ Geo-Activity: SE page 316
____ Examples: 1–4, SE pages 317–318
____ Extra Examples: TE pages 317–318; Internet Help at classzone.com
____ Checkpoint Exercises: 1–7, SE pages 317, 319
____ Technology Activity: SE page 324
____ Technology Keystrokes for Activity on SE page 324: CRB pages 27–28
____ Concept Check: TE page 319
____ Guided Practice Exercises: 1–7, SE page 320
____ Visualize It! Transparencies: 28

## APPLY/HOMEWORK

**Homework Assignment**

____ Block Schedule: pp. 320–323 Exs. 8–23, 25–44, Quiz 1

**Reteaching the Lesson**

____ Practice Masters: CRB pages 29–30 (Level A, Level B)
____ Reteaching with Practice: CRB pages 31–32 or Practice Workbook with Examples; Resources in Spanish

**Extending the Lesson**

____ Challenge: SE page 321; classzone.com

## ASSESSMENT OPTIONS

____ Daily Quiz (6.3): TE page 323, CRB page 37, or Transparencies
____ Standardized Test Practice: SE page 322; Transparencies
____ Quiz (6.1–6.3): SE page 323; CRB page 34

Notes _____

_____

_____

| CHAPTER PACING GUIDE | |
|---|---|
| **Day** | **Lesson** |
| 1 | 6.1 |
| 2 | 6.2 |
| 3 | **6.3** |
| 4 | 6.4 |
| 5 | 6.5 |
| 6 | 6.6 |
| 7 | Ch. 6 Review and Assess |

*Lesson 6.3*

# WARM-UP EXERCISES

For use before Lesson 6.3, pages 316–324

## State the theorem that justifies the statement.

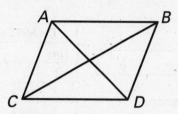

**1.** If $ABDC$ is a parallelogram, then $\overline{AB} \cong \overline{DC}$ and $\overline{AC} \cong \overline{BD}$.

**2.** If $ABDC$ is a parallelogram, then $\overline{AD}$ bisects $\overline{BC}$.

**3.** If $ABDC$ is a parallelogram, then $\angle A \cong \angle D$ and $\angle B \cong \angle C$.

**4.** If $ABDC$ is a parallelogram, then $m\angle A + m\angle B = 180°$ and $m\angle B + m\angle D = 180°$.

# DAILY HOMEWORK QUIZ

For use after Lesson 6.2, pages 309–315

## JKLM is a parallelogram.

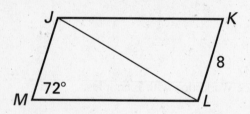

**1.** Find $JM$.

**2.** Find the measure of $\angle K$.

**3.** Find the measure of $\angle MJK$.

**4.** If $ML = 14$, find $JK$.

**5.** Suppose $\overline{MK}$ is added to the diagram, with the point of intersection with $\overline{JL}$ labeled as point $Q$. What can you conclude about $\overline{JQ}$ and $\overline{LQ}$?

**Geometry**
Chapter 6 Resource Book

# Technology Keystrokes

**For use with Technology Activity 6.3, page 324**

## TI-92

### *Explore*

**1.** Draw quadrilateral *ABCD*. [F2] 5 (Move cursor to desired location.) [ENTER] *A*
(Move cursor to desired location.) [ENTER] *B* [ENTER]
(Move cursor to desired location.) [ENTER] *C* [ENTER]
(Move cursor to desired location.) [ENTER] *D* [ENTER]
(Move cursor to point *A*.) [ENTER]

**2.** Measure the angles of the quadrilateral. First measure $\angle A$.
[F6] 3 (Move cursor to point *D*.) [ENTER]
(Move cursor to point *A*.) [ENTER] (Move cursor to point *B*.) [ENTER]
Repeat this procedure to measure $\angle B$, $\angle C$, and $\angle D$.

**3.** Drag the vertices. Use the [hand] key and the cursor pad to drag the vertices.

### *Think About It*

**1.** Find the slopes of $\overline{AB}$, $\overline{BC}$, $\overline{CD}$, and $\overline{DA}$. [F6] 4 (Move cursor to $\overline{AB}$.) [ENTER]
(Move cursor to $\overline{BC}$.) [ENTER] (Move cursor to $\overline{CD}$). [ENTER]
(Move cursor to $\overline{DA}$.) [ENTER]

**2–4.** These exercises do not require a graphing calculator.

**5.** Draw quadrilateral *EFGH*. [F2] 5 (Move cursor to desired location.) [ENTER] *E*
(Move cursor to desired location.) [ENTER] *F* [ENTER]
(Move cursor to desired location.) [ENTER] *G* [ENTER]
(Move cursor to desired location.) [ENTER] *H* [ENTER]
(Move cursor to point *E*.) [ENTER]
Draw $\overline{EG}$ and $\overline{FH}$. [F2] 5 (Move cursor to point *E*.) [ENTER]
(Move cursor to point *G*.) [ENTER] (Move cursor to point *F*.) [ENTER]
(Move cursor to point *H*.) [ENTER]
Draw the intersection of $\overline{EG}$ and $\overline{FH}$. [F2] 3 (Move cursor to $\overline{EG}$.) [ENTER]
(Move cursor to $\overline{FH}$.) [ENTER] *I*
Draw $\overline{EI}$, $\overline{IG}$, $\overline{FI}$, and $\overline{IH}$. [F2] 5 (Move cursor to point *E*.) [ENTER]
(Move cursor to point *I*.) [ENTER] [ENTER] (Move cursor to point *G*.) [ENTER]
(Move cursor to point *F*.) [ENTER] (Move cursor to point *I*.) [ENTER] [ENTER]
(Move cursor to point *H*.) [ENTER]
Measure $\overline{EI}$, $\overline{IG}$, $\overline{FI}$, and $\overline{IH}$. [F6] 1 (Move cursor to $\overline{EI}$.) [ENTER]
(Move cursor to $\overline{IG}$.) [ENTER] (Move cursor to $\overline{FI}$.) [ENTER]
(Move cursor to $\overline{IH}$.) [ENTER]
Drag the vertices. [F1] 1 Use the [hand] key and the cursor pad to drag the vertices.

NAME _____ DATE _____

# Technology Keystrokes
**For use with Technology Activity 6.3, page 324**

## SKETCHPAD

### Explore

1. Draw quadrilateral *ABCD*. Choose the segment tool. Draw $\overline{AB}$, $\overline{BC}$, $\overline{CD}$, and $\overline{DA}$ to form quadrilateral *ABCD*.

2. Measure ∠*A*, ∠*B*, ∠*C*, and ∠*D*. To measure ∠*A*, choose the selection arrow tool. Select point *D*. Hold down the shift key and select points *A* and *B*. Choose **Angle** from the **Measure** menu. Click on a blank area of the window to deselect all objects. Repeat this procedure to measure ∠*B*, ∠*C*, and ∠*D*.

3. Drag the vertices. Choose the selection arrow tool. Select any vertex and drag.

### Think About It

1. Find the slopes of $\overline{AB}$, $\overline{BC}$, $\overline{CD}$, and $\overline{DA}$. Choose the selection arrow tool. Select $\overline{AB}$. Hold down the shift key and select $\overline{BC}$, $\overline{CD}$, and $\overline{DA}$. Choose **Slope** from the **Measure** menu.

2–4. No geometry software is required.

5. Draw quadrilateral *EFGH*. Choose the segment tool. Draw $\overline{EF}$, $\overline{FG}$, $\overline{GH}$, and $\overline{HE}$ to form quadrilateral *EFGH*.

   Draw $\overline{EG}$ and $\overline{FH}$. Choose the segment tool. Draw $\overline{EG}$ and $\overline{FH}$.

   Draw the intersection of $\overline{EG}$ and $\overline{FH}$. Choose the selection arrow tool. Select $\overline{EG}$. Hold down the shift key and select $\overline{FH}$. Choose **Point At Intersection** from the **Construct** menu.

   Measure $\overline{EI}$, $\overline{IG}$, $\overline{FI}$, and $\overline{IH}$. Choose the segment tool. Draw $\overline{EI}$, $\overline{IG}$, $\overline{FI}$, and $\overline{IH}$. Choose the selection arrow tool. Select $\overline{EI}$. Hold down the shift key and select $\overline{IG}$, $\overline{FI}$, and $\overline{IH}$. Choose **Length** from the **Measure** menu.

   Drag the vertices. Choose the selection arrow tool. Select and drag any of the vertices.

## *Practice A*
**For use with pages 316–324**

**Match the figure with the method you would use to show that it is a parallelogram.**

**A.** $x° + y° = 180°$

**B.**

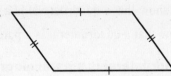

**C.**

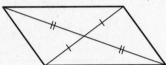

**D.**

1. Show that both pairs of opposite sides are congruent.

2. Show that both pairs of opposite angles are congruent.

3. Show that one angle is supplementary with both of its consecutive angles.

4. Show that the diagonals bisect each other.

**Tell whether the quadrilateral is a parallelogram.**
**Explain your reasoning.**

**5.**

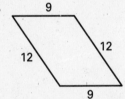

**6.**

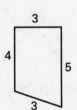

**7.**

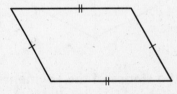

**8.**

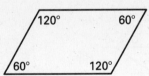

**9.**

**10.**

**11.**

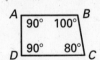

**12.**

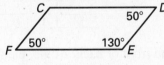

**13.**

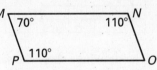

**14.**

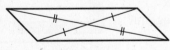

**15.**

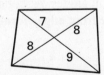

**16.**

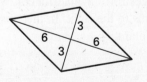

**The figure at the right shows a musical note with quadrilateral *ABCD*.**

17. Tell two ways to show that quadrilateral *ABCD* is a parallelogram using angles.

18. Tell two ways to show that quadrilateral *ABCD* is a parallelogram using sides.

NAME _____ DATE _____

# Practice B

**For use with pages 316–324**

1. State two ways to show that a quadrilateral is a parallelogram using opposite sides.

2. State two ways to show that a quadrilateral is a parallelogram using angles.

3. State a way to show that a quadrilateral is a parallelogram using diagonals.

**Tell whether the quadrilateral is a parallelogram. Explain your reasoning.**

4.
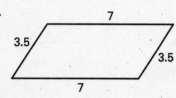

5.
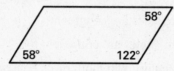

6.
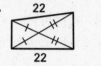

7.
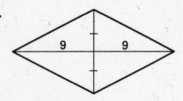

8.
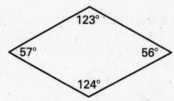

9.

10.
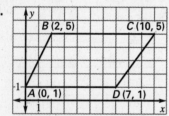

11.
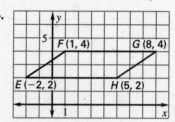

12.

**Use the slopes of the segments in the diagram to determine if the quadrilateral is a parallelogram.**

13.
```
      y
   B (2, 5)        C (10, 5)

-1
   A (0, 1)      D (7, 1)
    1                    x
```

14.
```
        y
   5  F (1, 4)      G (8, 4)

 E (−2, 2)      H (5, 2)
        1            x
```

15. The drawings show the same box with no front or back. At top, the front edges form a quadrilateral with four right angles. At bottom, the box leans to the right. In this position do the front edges form a parallelogram? Explain.

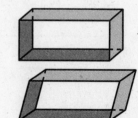

NAME _____ DATE _____

# *Reteaching with Practice*

**For use with pages 316–324**

**GOAL** **Show that a quadrilateral is a parallelogram.**

---

**VOCABULARY**

**Theorem 6.6**
If both pairs of opposite sides of a quadrilateral are congruent, then the quadrilateral is a parallelogram.

**Theorem 6.7**
If both pairs of opposite angles of a quadrilateral are congruent, then the quadrilateral is a parallelogram.

**Theorem 6.8**
If an angle of a quadrilateral is supplementary to both of its consecutive angles, then the quadrilateral is a parallelogram.

**Theorem 6.9**
If the diagonals of a quadrilateral bisect each other, then the quadrilateral is a parallelogram.

---

**EXAMPLE 1** *Use Opposite Sides*

Tell whether the quadrilateral is a parallelogram. Explain your reasoning.

**a.**     **b.**

**SOLUTION**

**a.** The quadrilateral is a parallelogram because both pairs of opposite sides are congruent.

**b.** The quadrilateral is not a parallelogram. Both pairs of opposite sides are not congruent.

**Exercises for Example 1**

**Tell whether the quadrilateral is a parallelogram. Explain your reasoning.**

**1.** 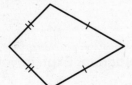    **2.** 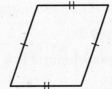    **3.**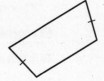

---

NAME_____ DATE _____

# Reteaching with Practice

**For use with pages 316–324**

**EXAMPLE 2** **Use Opposite Angles**

Tell whether the quadrilateral is a parallelogram. Explain your reasoning.

**a.**

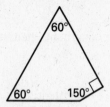

**b.**

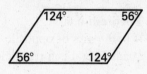

### SOLUTION

**a.** The quadrilateral is not a parallelogram. Both pairs of opposite angles are not congruent.

**b.** The quadrilateral is a parallelogram because both pairs of opposite angles are congruent.

## Exercises for Example 2

**Tell whether the quadrilateral is a parallelogram. Explain your reasoning.**

**4.**          **5.**          **6.**

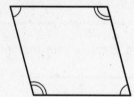

**EXAMPLE 3** **Use Diagonals and Consecutive Angles**

Tell whether the quadrilateral is a parallelogram. Explain your reasoning.

**a.**

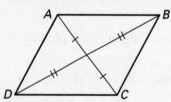

**b.**

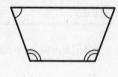

### SOLUTION

**a.** The diagonals of *ABCD* bisect each other. So, by Theorem 6.9, *ABCD* is a parallelogram.

**b.** ∠*H* is supplementary to ∠*E* and ∠*G* (87° + 93° = 180°). So, by Theorem 6.8, *EFGH* is a parallelogram.

## Exercises for Example 3

**Tell whether the quadrilateral is a parallelogram. Explain your reasoning.**

**7.**          **8.**          **9.**

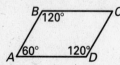

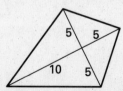

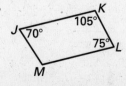

NAME _____ DATE _____

# Quick Catch-Up for Absent Students

**For use with pages 316–324**

The items checked below were covered in class on (date missed) _____

**Lesson 6.3: Showing Quadrilaterals are Parallelograms (pp. 316–319)**

____ **Goal:** Show that a quadrilateral is a parallelogram.

*Material Covered:*

____ Geo-Activity: Making Parallelograms

____ Student Help: Study Tip

____ Example 1: Use Opposite Sides

____ Example 2: Use Opposite Angles

____ Example 3: Use Consecutive Angles

____ Example 4: Use Diagonals

**Technology Activity 6.3: Making Parallelograms (p. 324)**

____ **Goal:** Use angle measures to show that a quadrilateral is a parallogram.

____ Student Help: Study Tip

Other (specify) _____

_____

**Homework and Additional Learning Support**

____ Textbook exercises (teacher to specify) pp. 320–323 _____

_____

____ Internet: More Examples at classzone.com

____ *Reteaching with Practice* worksheet

NAME _____ DATE _____

## Quiz 1

**For use after Lessons 6.1–6.3**

**1.** What type of polygon has six sides?

**Find the measure of ∠A.**

**2.**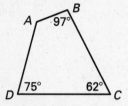

**3.**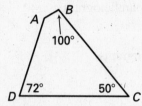

**In Exercises 4–7, use the diagram.** *ABCD* **is a parallelogram and** *m∠DAB* = 115°.

**4.** Find *m∠ABC*.

**5.** Find *m∠BCD*.

**6.** Find *CD*.

**7.** Find *BD*.

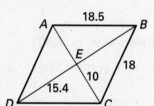

**Decide whether you are given enough information to show that the quadrilateral is a parallelogram. Explain your reasoning.**

**8.**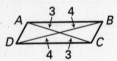

**9.**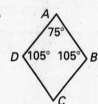

**10.**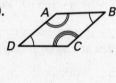

TEACHER'S NAME _____ CLASS _____ ROOM _____ DATE _____

## *Lesson Plan*

2-day lesson (See *Pacing the Chapter,* TE page 300A)    **For use with pages 325–330**

**GOAL**  **Use properties of special types of parallelograms.**

State/Local Objectives _____

_____

### ✓ Check the items you wish to use for this lesson.

**STARTING OPTIONS**
____ Homework Check (6.3): TE page 320; Answer Transparencies
____ Homework Quiz (6.3): TE page 323, CRB page 37, or Transparencies
____ Warm-Up: CRB page 37 or Transparencies

**TEACHING OPTIONS**
____ Examples: Day 1: 1–2, SE pages 325–326; Day 2: 3–4, SE page 327
____ Extra Examples: TE pages 326–327; Internet Help at classzone.com
____ Checkpoint Exercises: Day 1: 1–3, SE pages 325–326; Day 2: 4–6, SE page 328
____ Technology Activity with Keystrokes: CRB pages 38–39
____ Concept Check: TE page 327
____ Guided Practice Exercises: Day 1: 1, SE page 328; Day 2: 2–6, SE page 328
____ Visualize It! Transparencies: 29

**APPLY/HOMEWORK**
**Homework Assignment**
____ Basic: Day 1: pp. 328–330 Exs. 7–12, 14, 15, 18, 19, 27, 28
     Day 2: pp. 328–330 Exs. 13, 16, 17, 20–22, 29–34
____ Average: Day 1: pp. 328–330 Exs. 7–12, 14, 15, 18, 19, 27, 28
     Day 2: pp. 328–330 Exs. 13, 16, 17, 20–22, 29–34
____ Advanced: Day 1: pp. 328–330 Exs. 7–12, 14, 15, 18, 19, 27–31
     Day 2: pp. 328–330 Exs. 13, 16, 17, 20–22, 23–26*, 32–34; EC: classzone.com

**Reteaching the Lesson**
____ Practice Masters: CRB pages 40–41 (Level A, Level B)
____ Reteaching with Practice: CRB pages 42–43 or Practice Workbook with Examples;
     Resources in Spanish

**Extending the Lesson**
____ Real-Life Application: CRB page 45
____ Challenge: SE page 330; classzone.com

**ASSESSMENT OPTIONS**
____ Daily Quiz (6.4): TE page 330, CRB page 48, or Transparencies
____ Standardized Test Practice: SE page 330; Transparencies

Notes _____

_____

_____

_____

TEACHER'S NAME _____ CLASS _____ ROOM _____ DATE _____

# Lesson Plan for Block Scheduling

**1-block lesson (See *Pacing the Chapter*, TE page 300A)**          **For use with pages 325–330**

**GOAL**   **Use properties of special types of parallelograms.**

State/Local Objectives _____

_____

_____

✓ **Check the items you wish to use for this lesson.**

## STARTING OPTIONS
____ Homework Check (6.3): TE page 320; Answer Transparencies
____ Homework Quiz (6.3): TE page 323, CRB page 37,
      or Transparencies
____ Warm-Up: CRB page 37 or Transparencies

## TEACHING OPTIONS
____ Examples: 1–4, SE pages 325–327
____ Extra Examples: TE pages 326–327; Internet Help at classzone.com
____ Checkpoint Exercises: 1–6, SE pages 325–326, 328
____ Technology Activity with Keystrokes: CRB pages 38–39
____ Concept Check: TE page 327
____ Guided Practice Exercises: 1–6, SE page 328
____ Visualize It! Transparencies: 29

## APPLY/HOMEWORK
**Homework Assignment**
____ Block Schedule: pp. 328–330 Exs. 7–22, 27–34

**Reteaching the Lesson**
____ Practice Masters: CRB pages 40–41 (Level A, Level B)
____ Reteaching with Practice: CRB pages 42–43 or Practice Workbook with Examples;
      Resources in Spanish

**Extending the Lesson**
____ Real-Life Application: CRB page 45
____ Challenge: SE page 330; classzone.com

## ASSESSMENT OPTIONS
____ Daily Quiz (6.4): TE page 330, CRB page 48, or Transparencies
____ Standardized Test Practice: SE page 330; Transparencies

Notes _____

_____

_____

| CHAPTER PACING GUIDE | |
|---|---|
| **Day** | **Lesson** |
| 1 | 6.1 |
| 2 | 6.2 |
| 3 | 6.3 |
| 4 | **6.4** |
| 5 | 6.5 |
| 6 | 6.6 |
| 7 | Ch. 6 Review and Assess |

NAME _____ DATE _____

# WARM-UP EXERCISES

For use before Lesson 6.4, pages 325–330

**WXYZ is a parallelogram. Find the value of x.**

**1.**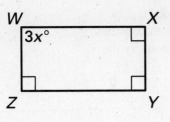

**2.**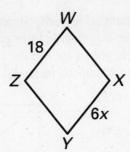

- - - - - - - - - - - - - - - - - - - - - - - - - - - - - - - - - - - - - - -

# DAILY HOMEWORK QUIZ

For use after Lesson 6.3, pages 316–324

## Tell whether the quadrilateral is a parallelogram. Explain your reasoning.

**1.**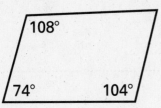

**2.**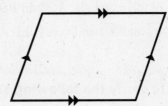

**3.**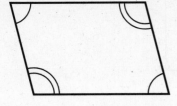

**4.**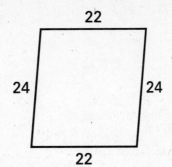

# Technology Activity

**For use with pages 325–330**

**GOAL** **Use geometry software to verify statements about special parallelograms.**

Geometry software can be used to verify statements about special parallelograms. For example, you could use geometry software to construct the rectangle below. Then, you could use the software's tools to verify the statement about the diagonals of the rectangle.

Given rectangle *ABCD*, verify that $\overline{AC} \cong \overline{BD}$.

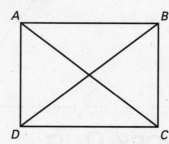

## Activity

❶ Construct rectangle *ABCD* (see figure above). Use the software's grid feature to ensure that you construct the sides such that the opposite sides are parallel and congruent.

❷ Construct the diagonals of the rectangle, $\overline{AC}$ and $\overline{BD}$.

❸ Measure the lengths of $\overline{AC}$ and $\overline{BD}$ and verify that $\overline{AC} \cong \overline{BD}$.

## Exercises

**Use geometry software to verify the following statements.**

1. Construct square *ABCD*. Verify that *ABCD* is both a rectangle and a rhombus.

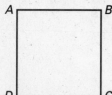

2. Construct square *ABCD*. Construct diagonals $\overline{AC}$ and $\overline{BD}$. Verify that $\overline{AC} \perp \overline{BD}$.

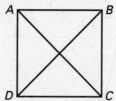

3. Construct square *ABCD*. Construct diagonals $\overline{AC}$ and $\overline{BD}$. Verify that $\overline{AC}$ bisects $\angle DAB$ and $\angle BCD$ and $\overline{BD}$ bisects $\angle ADC$ and $\angle ABC$.

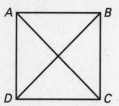

# Technology Keystrokes

**For use with Technology Activity, CRB page 38**

## TI-92

1. Construct rectangle *ABCD*.

   **F8** 9 (Set Coordinate Axes to RECTANGULAR and Grid to ON.) **ENTER**

   **F3** 4 (Move cursor to point (−2, 1) and prompt says, "ON THIS POINT OF THE GRID.") **ENTER** *A* (Move cursor to point (2, 1) and prompt says, "ON THIS POINT OF THE GRID.") **ENTER** *B* (Move cursor to point (2, −1) and prompt says, "ON THIS POINT OF THE GRID.") **ENTER** *C* (Move cursor to point (−2, −1) and prompt says, "ON THIS POINT OF THE GRID.") **ENTER** *D* (Move cursor to point *A*.) **ENTER**

2. Construct the diagonals of the rectangle, $\overline{AC}$ and $\overline{BD}$.

   **F2** 5 (Move cursor to point *A*.) **ENTER**
   (Move cursor to point *C*.) **ENTER** (Move cursor to point *B*.) **ENTER**
   (Move cursor to point *D*.) **ENTER**

3. Measure the lengths of $\overline{AC}$ and $\overline{BD}$.

   **F6** 1 (Move cursor to $\overline{AC}$.) **ENTER** (Move cursor to $\overline{BD}$.) **ENTER**

## SKETCHPAD

1. Turn on the axes and the grid. Choose **Snap To Grid** from the **Graph** menu.

   Choose the segment tool. Draw a segment from (−2, 1) to (2, 1). Draw a segment from (2, 1) to (2, −1). Draw a segment from (2, −1) to (−2, −1). Draw a segment from (−2, −1) to (−2, 1).

   Label the points. Choose the text tool. Double click on the label for (−2, 1), type *A*, and click **OK**. Double click on the label for (2, 1), type *B*, and click **OK**. Double click on the label for (2, −1), type *C*, and click **OK**. Double click on the label for (−2, −1), type *D*, and click **OK**.

2. Construct the rectangle's diagonals. Choose the segment tool. Draw $\overline{AC}$ and $\overline{BD}$.

3. Measure the lengths of $\overline{AC}$ and $\overline{BD}$. Choose the selection arrow tool. Select $\overline{AC}$. Hold down the shift key and select $\overline{BD}$. Choose **Length** from the **Measure** menu.

## Practice A
For use with pages 325–330

**Complete the statement.**

1. If a quadrilateral has four congruent sides, then it is a __?__.

2. If a quadrilateral has four right angles, then it is a __?__.

3. If a quadrilateral has four congruent sides and four right angles, then it is a __?__.

4. The diagonals of a rhombus are __?__.

5. The diagonals of a rectangle are __?__.

**Use the diagram to complete the statement.**

6. rectangle *WXYZ*

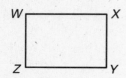

   a. $m\angle W = $ __?__
   b. $m\angle X = $ __?__
   c. $m\angle Y = $ __?__
   d. $m\angle Z = $ __?__

7. rhombus *DEFG*

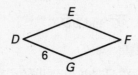

   a. $DE = $ __?__
   b. $EF = $ __?__
   c. $FG = $ __?__

8. square *NRPQ*

   a. $m\angle R = $ __?__
   b. $PQ = $ __?__
   c. $QN = $ __?__
   d. $m\angle N = $ __?__

**Name the special quadrilateral using the information in the diagram.**

9.

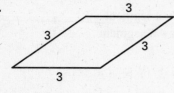

10.

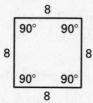

11.

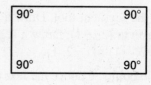

**Find the value of *x*.**

12. square *EFGH*

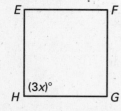

13. rectangle *KLMN*
    $KM = 3x + 2$
    $NL = 4x$

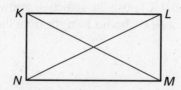

14. rhombus *PQRS*

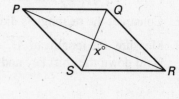

**The figure at the right shows a portion of a pasture fence with parallelogram *ABCD*.**

15. If $AB = BC = CD = DA$, name the special quadrilateral *ABCD*.

16. If $m\angle A = m\angle B = m\angle C = m\angle D = 90°$, name the special quadrilateral *ABCD*.

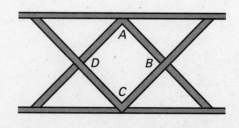

NAME_____ DATE _____

# Practice B
### For use with pages 325–330

**Write each theorem/corollary statement using symbols for quadrilateral *ABCD*.**

1. If a quadrilateral has four congruent sides, then it is a rhombus.

2. If a quadrilateral has four right angles, then it is a rectangle.

3. If a quadrilateral has four congruent sides and four right angles, then it is a square.

4. The diagonals of a rhombus are perpendicular.

5. The diagonals of a rectangle are congruent.

**Find the measures.**

6. rhombus *TUVW*

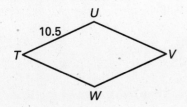

   **a.** $UV = $ __?__
   **b.** $VW = $ __?__
   **c.** $WT = $ __?__

7. rectangle *QRST*

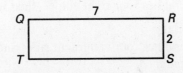

   **a.** $m\angle Q = $ __?__
   **b.** $TS = $ __?__
   **c.** $QT = $ __?__

8. square *BCDE*

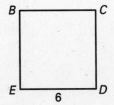

   **a.** $m\angle C = $ __?__
   **b.** $BC = $ __?__
   **c.** $CD = $ __?__

**List each quadrilateral for which the statement is true.**

 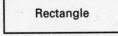 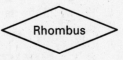

9. Opposite angles are congruent.

10. Diagonals bisect each other.

11. It has four congruent sides and four right angles.

12. It has four right angles.

**Find the value of the variable.**

13. rectangle *KLMN*

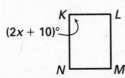

14. rhombus *ABCD*

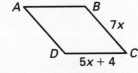

15. square *RSTU*

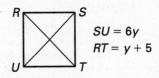

$SU = 6y$
$RT = y + 5$

16. You want to make a diamond-shaped frame for a wall hanging out of a 52-inch bamboo stick. If the diamond is to be a square, how long can the sides be? What should be true about the distances between the opposite corners?

NAME_____  DATE _____

# *Reteaching with Practice*

For use with pages 325–330

**GOAL** Use properties of special types of parallelograms.

---

**VOCABULARY**

A **rhombus** is a parallelogram with four congruent sides.

A **rectangle** is a parallelogram with four right angles.

A **square** is a parallelogram with four congruent sides and four right angles.

**Rhombus Corollary**
If a quadrilateral has four congruent sides, then it is a rhombus.

**Rectangle Corollary**
If a quadrilateral has four right angles, then it is a rectangle.

**Square Corollary**
If a quadrilateral has four congruent sides and four right angles, then it is a square.

**Theorem 6.10**
The diagonals of a rhombus are perpendicular.

**Theorem 6.11**
The diagonals of a rectangle are congruent.

---

**EXAMPLE 1** *Use Properties of Special Parallelograms*

In the diagram, *ABCD* is a square.

Find the values of *x* and *y*.

**SOLUTION**

By definition, a square has four right angles.

$m\angle A = 90°$       Definition of a square

$10x° = 90°$       Substitute $10x°$ for $m\angle A$.

$x = 9$       Divide each side by 10.

By definition, a square has four congruent sides. So, $AB = BC$.

$5 = y - 3$       Substitute 5 for *AB* and $y - 3$ for *BC*.

$8 = y$       Add 3 to each side.

---

**Exercises for Example 1**
..........................................................................................

**Find the values of the variables.**

**1.** rectangle *ABCD*

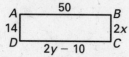

**2.** rhombus *EFGH*

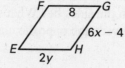

**3.** square *JKLM*

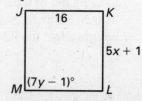

NAME _____ DATE _____

# Reteaching with Practice

For use with pages 325–330

---

**EXAMPLE 2**  **Use Diagonals of a Rhombus**

*JKLM* is a rhombus.
Find the value of *x*.

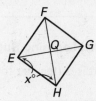

### SOLUTION

By Theorem 6.10, the diagonals of a rhombus are perpendicular.
Therefore, $\angle KNL$ is a right angle, so $\triangle KNL$ is a right triangle.

By the Corollary to the Triangle Sum Theorem, the acute angles
of a right triangle are complementary.

| | |
|---|---|
| $m\angle NKL + m\angle KLN = 90°$ | Corollary to the Triangle Sum Theorem |
| $55° + 7x° = 90°$ | Substitute 55° for $m\angle NKL$ and 7x° for $m\angle KLN$. |
| $7x = 35$ | Subtract 55 from each side. |
| $x = 5$ | Divide each side by 7. |

### Exercises for Example 2

**Find the value of *x*.**

**4.** rhombus *ABCD*

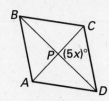

**5.** rhombus *EFGH*

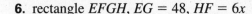

---

**EXAMPLE 3**  **Use Diagonals of a Rectangle**

*ABCD* is a rectangle. $AC = 16$.
$BD = 5x + 1$. Find the value of *x*.

### SOLUTION

By Theorem 6.11, the diagonals of a rectangle are congruent.
Therefore, $AC = BD$.

| | |
|---|---|
| $16 = 5x + 1$ | Substitute 16 for $AC$ and $5x + 1$ for $BD$. |
| $15 = 5x$ | Subtract 1 from each side. |
| $3 = x$ | Divide each side by 5. |

### Exercises for Example 3

**Find the value of *x*.**

**6.** rectangle *EFGH*, $EG = 48$, $HF = 6x$

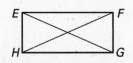

**7.** rectangle *WXYZ*, $XZ = 37$, $WY = 5x + 2$

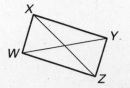

# *Quick Catch-Up for Absent Students*

**For use with pages 325–330**

The items checked below were covered in class on (date missed) _____

**Lesson 6.4: Rhombuses, Rectangles, and Squares (pp. 325–328)**

_____ **Goal:** Use properties of special types of parallelograms.

*Material Covered:*

_____ Example 1: Use Properties of Special Parallelograms

_____ Student Help: Study Tip

_____ Example 2: Identify Special Quadrilaterals

_____ Student Help: Look Back

_____ Example 3: Use Diagonals of a Rhombus

_____ Example 4: Use Diagonals of a Rectangle

*Vocabulary:*

rhombus, p. 325                                    square, p. 325
rectangle, p. 325

_____ Other (specify) _____

_____

**Homework and Additional Learning Support**

_____ Textbook exercises (teacher to specify) <u>pp. 328–330</u> _____

_____

_____ Internet: More Examples at classzone.com

_____ *Reteaching with Practice* worksheet

*Lesson 6.4*

NAME_____ DATE _____

# *Real-Life Application:*
# *When Will I Ever Use This?*

**For use with pages 325–330**

## Quilting

Although quilting can be traced to prehistoric times,
it flourished in Europe from the 17th through the 19th
centuries. Quilting was brought over to America with the
colonists, who often used the technique for such practical
purposes as clothing and bedcovers. At first the designs
imitated those of the English and Dutch, but soon an
American style quilt took form.

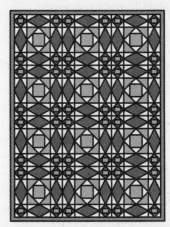

Using such techniques as patchwork and applique, quilts
commonly display geometric patterns or pictures of
animals, people, and objects in nature. Often quilts are
used to remember historical events, important family
memories, or stories. They have truly become a symbolic
representation of America's heritage.

More traditional quilts are fashioned by piecing together
quilt blocks. In the illustration of the Storm at Sea quilt
block below at the right, you can see how the repetition
of the pattern creates the finished product above at the right.

### In Exercises 1 and 2, refer to the quilt block below.

1. Use as many words as possible to describe each shape
   from the quilt block: *rhombus, rectangle, square.* You
   may wish to use a centimeter ruler and/or a protractor.

   **a.**  **b.**  **c.** ▢

2. Copy and complete the table by counting the number of each shape.

| *Shape* | Rhombus | Rectangle | Square |
|---------|---------|-----------|--------|
| *Number* | _____ | _____ | _____ |

TEACHER'S NAME _____ CLASS _____ ROOM _____ DATE _____

## *Lesson Plan*

2-day lesson (See *Pacing the Chapter,* TE page 300A)                    For use with pages 331–336

**GOAL**   Use properties of trapezoids.

State/Local Objectives _____

_____

### ✓ Check the items you wish to use for this lesson.

**STARTING OPTIONS**

____ Homework Check (6.4): TE page 328; Answer Transparencies

____ Homework Quiz (6.4): TE page 330, CRB page 48, or Transparencies

____ Warm-Up: CRB page 48 or Transparencies

**TEACHING OPTIONS**

____ Technology Activity: SE page 331

____ Technology Keystrokes for Activity on SE page 331: CRB pages 49–50

____ Examples: Day 1: 1, SE page 332; Day 2: 2, SE page 333

____ Extra Examples: TE page 333

____ Checkpoint Exercises: Day 1: 1–3, SE page 333; Day 2: 4–6, SE page 333

____ Concept Check: TE page 333

____ Guided Practice Exercises: Day 1: 1–5, SE page 334; Day 2: 6–8, SE page 334

____ Visualize It! Transparencies: 30

**APPLY/HOMEWORK**

**Homework Assignment**

____ Basic: Day 1: pp. 334–336 Exs. 14–19, 35, 39–46

Day 2: pp. 334–336 Exs. 9–13, 20–26, 34, 36–38

____ Average: Day 1: pp. 334–336 Exs. 14–19, 35–46

Day 2: pp. 334–336 Exs. 9–13, 20–32, 34

____ Advanced: Day 1: pp. 334–336 Exs. 14–19, 35–46

Day 2: pp. 334–336 Exs. 20–32, 33*, 34; EC: classzone.com

**Reteaching the Lesson**

____ Practice Masters: CRB pages 51–52 (Level A, Level B)

____ Reteaching with Practice: CRB pages 53–54 or Practice Workbook with Examples;
Resources in Spanish

**Extending the Lesson**

____ Challenge: SE page 336; classzone.com

**ASSESSMENT OPTIONS**

____ Daily Quiz (6.5): TE page 336, CRB page 58, or Transparencies

____ Standardized Test Practice: SE page 336; Transparencies

Notes _____

_____

_____

TEACHER'S NAME _____ CLASS _____ ROOM _____ DATE _____

# Lesson Plan for Block Scheduling

1-block lesson (See *Pacing the Chapter,* TE page 300A)          For use with pages 331–336

**GOAL**   Use properties of trapezoids.

State/Local Objectives _____

_____

_____

✓ **Check the items you wish to use for this lesson.**

**STARTING OPTIONS**
____ Homework Check (6.4): TE page 328; Answer Transparencies
____ Homework Quiz (6.4): TE page 330, CRB page 48,
     or Transparencies
____ Warm-Up: CRB page 48 or Transparencies

| CHAPTER PACING GUIDE | |
|---|---|
| **Day** | **Lesson** |
| 1 | 6.1 |
| 2 | 6.2 |
| 3 | 6.3 |
| 4 | 6.4 |
| 5 | **6.5** |
| 6 | 6.6 |
| 7 | Ch. 6 Review and Assess |

**TEACHING OPTIONS**
____ Technology Activity: SE page 331
____ Technology Keystrokes for Activity on SE page 331: CRB pages 49–50
____ Examples: 1–2, SE pages 332–333
____ Extra Examples: TE page 333
____ Checkpoint Exercises: 1–6, SE page 333
____ Concept Check: TE page 333
____ Guided Practice Exercises: 1–8, SE page 334
____ Visualize It! Transparencies: 30

**APPLY/HOMEWORK**
**Homework Assignment**
____ Block Schedule:  pp. 334–336 Exs. 9–32, 34–46

**Reteaching the Lesson**
____ Practice Masters: CRB pages 51–52  (Level A, Level B)
____ Reteaching with Practice: CRB pages 53–54 or Practice Workbook with Examples;
     Resources in Spanish

**Extending the Lesson**
____ Challenge: SE page 336; classzone.com

**ASSESSMENT OPTIONS**
____ Daily Quiz (6.5): TE page 336, CRB page 58, or Transparencies
____ Standardized Test Practice: SE page 336; Transparencies

Notes _____

_____

_____

NAME _____ DATE _____

# WARM-UP EXERCISES

For use before Lesson 6.5, pages 331–336

## Find the coordinates of the midpoint of $\overline{AB}$.

**1.** $A(0, 0)$, $B(4, -2)$          **2.** $A(3, -5)$, $B(-7, -1)$

## Solve for $y$.

**3.** $25 = \frac{1}{2}(16 + y)$          **4.** $26 = \frac{1}{2}(y + 47)$

# DAILY HOMEWORK QUIZ

For use after Lesson 6.4, pages 325–330

## Use the information in the diagram to name the special quadrilateral.

**1.**

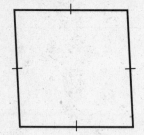

**2.**

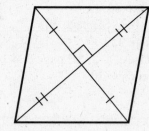

**3.**

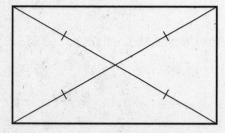

**4.**

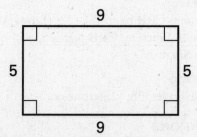

**5.** Kim arranges four metersticks to make a parallelogram.
Then she adjusts the metersticks so that each pair meets
at a right angle. What shape has Kim formed?

# Technology Keystrokes

**For use with Technology Activity 6.5, page 331**

## TI-92

### *Explore*

**1.** Draw $\overleftrightarrow{AB}$. `F2` 4 (Move cursor to desired location.) `ENTER`

(Move cursor to desired location.) `ENTER`

Draw points $A$ and $B$ on the line. `F2` 1 (Move cursor to desired location on the

line.) `ENTER` $A$ (Move cursor to desired location on the line.) `ENTER` $B$

Draw point $C$ not on $\overleftrightarrow{AB}$. (Move cursor to desired location.) `ENTER` $C$

Construct a line parallel to $\overleftrightarrow{AB}$ through point $C$.

`F4` 2 (Move cursor to point $C$.) `ENTER` (Move cursor to $\overleftrightarrow{AB}$.) `ENTER`

**2.** Draw point $D$. `F2` 1 (Move cursor to desired location.) `ENTER` $D$

Draw $\overline{AD}$ and $\overline{BC}$. `F2` 5 (Move cursor to point $A$.) `ENTER`

(Move cursor to point $D$.) `ENTER` (Move cursor to point $B$.) `ENTER`

(Move cursor to point $C$.) `ENTER`

**3.** Construct and label the midpoints of $\overline{AD}$ and $\overline{BC}$.

`F4` 3 (Move cursor to $\overline{AD}$.) `ENTER` $E$ (Move cursor to $\overline{BC}$.) `ENTER` $F$

Draw $\overline{EF}$. `F2` 5 (Move cursor to point $E$.) `ENTER`

(Move cursor to point $F$.) `ENTER`

### *Think About It*

**1.** Draw $\overline{AB}$ and $\overline{DC}$. `F2` 5 (Move cursor to point $A$.) `ENTER` (Move cursor to

point $B$.) `ENTER` (Move cursor to point $D$.) `ENTER`

(Move cursor to point $C$.) `ENTER`

Measure $\overline{AB}$, $\overline{DC}$, and $\overline{EF}$. `F6` 1 (Move cursor to $\overline{AB}$. The prompt will ask

"WHICH OBJECT?" `ENTER` Choose "THIS SEGMENT.") `ENTER`

(Move cursor to $\overline{DC}$. The prompt will ask "WHICH OBJECT?" `ENTER`

Choose "THIS SEGMENT.") `ENTER` (Move cursor to $\overline{EF}$.) `ENTER`

**2.** Calculate $\frac{(AB + DC)}{2}$. `F6` 6 `(` (Use cursor to highlight $AB$.) `ENTER` `+`

(Use cursor to highlight $DC$.) `ENTER` `)` `÷` 2 `ENTER`

**3.** Move points $A$, $B$, C, and $D$. `F1` 1 Use the ☞ key and the cursor pad to drag the points.

**4.** Measure the slopes of $\overleftrightarrow{AB}$, $\overleftrightarrow{DC}$, and $\overline{EF}$. `F6` 4 (Move cursor to $\overleftrightarrow{AB}$.) `ENTER`

(Choose "THIS LINE.") (Move cursor to $\overleftrightarrow{DC}$.) `ENTER` (Choose "THIS LINE.")

(Move cursor to $\overline{EF}$.) `ENTER`

# Technology Keystrokes

For use with Technology Activity 6.5, page 331

## SKETCHPAD

### Explore

1. Draw $\overleftrightarrow{AB}$. Choose the line tool. Draw $\overleftrightarrow{AB}$.

   Draw point $C$ not on $\overleftrightarrow{AB}$. Choose the point tool. Draw point $C$.

   Construct a line parallel to $\overleftrightarrow{AB}$ through point $C$. Choose the selection arrow tool.
   Select $\overleftrightarrow{AB}$. Hold down the shift key and select point $C$. Choose **Parallel Line** from the
   **Construct** menu.

2. Draw point $D$. Choose the point tool. Draw point $D$.

   Draw $\overline{AD}$ and $\overline{BC}$. Choose the segment tool. Draw $\overline{AD}$ and $\overline{BC}$.

3. Construct and label the midpoints of $\overline{AD}$ and $\overline{BC}$. Choose the selection arrow tool.
   Select $\overline{BC}$. Hold down the shift key and select $\overline{AD}$ . Choose **Point At Midpoint** from the
   **Construct** menu.

   Draw $\overline{EF}$. Choose the segment tool. Draw $\overline{EF}$.

### Think About It

1. Draw $\overline{AB}$ and $\overline{DC}$. Choose the segment tool. Draw $\overline{AB}$ and $\overline{DC}$.

   Measure $\overline{AB}$, $\overline{DC}$, and $\overline{EF}$. Choose the selection arrow tool. Select $\overline{AB}$. Hold down
   the shift key and select $\overline{DC}$ and $\overline{EF}$. Choose **Length** from the **Measure** menu.

2. Calculate $\frac{(AB + DC)}{2}$. Choose **Calculate** from the **Measure** menu. Click  **(** ,

   $m\overline{AB}$,  **+** , $m\overline{DC}$,  **)** ,  **/** , and 2. Click  **OK** .

3. Move points $A$, $B$, $C$, and $D$. Choose the selection arrow tool. Click and drag the points.

4. Measure the slopes of $\overleftrightarrow{AB}$, $\overrightarrow{DC}$, and $\overline{EF}$. Choose the selection arrow tool. Select $\overleftrightarrow{AB}$.
   Hold down the shift key and select $\overrightarrow{DC}$ and $\overline{EF}$. Choose **Slope** from the **Measure** menu.

NAME_____ DATE _____

# Practice A

For use with pages 331–336

**Match the key words with the descriptive phrase.**

1. trapezoid          **A.** a segment that connects the midpoints of the legs of a trapezoid

2. bases of a trapezoid     **B.** a quadrilateral with exactly one pair of parallel sides

3. legs of a trapezoid      **C.** the parallel sides of a trapezoid

4. midsegment of a trapezoid  **D.** a trapezoid that has congruent legs

5. isosceles trapezoid      **E.** the nonparallel sides of a trapezoid

**Decide whether the quadrilateral is a *trapezoid,* an *isosceles trapezoid,* or *neither*.**

6. 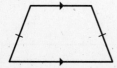    7.     8.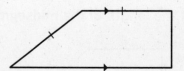

***CDEF* is an isosceles trapezoid. Find the missing angle measures.**

9. 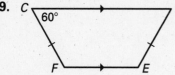    10. 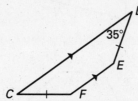    11.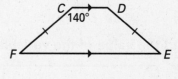

***EFGH* is a trapezoid. Find the missing angle measures.**

12. 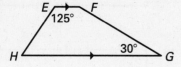    13. 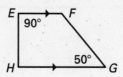    14.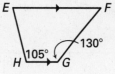

**Find the length of the midsegment $\overline{MN}$ of the trapezoid.**

15. 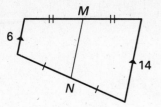    16. 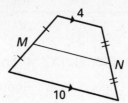    17.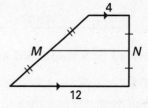

NAME_____ DATE _____

# Practice B

For use with pages 331–336

**Supply the missing word to complete the statement.**

1. If a trapezoid is isosceles, then each pair of base angles is __?__ .

2. If a trapezoid has a pair of congruent __?__ angles, then it is isosceles.

3. The length of the __?__ of a trapezoid is half the sum of the lengths of the bases.

4. A trapezoid is a quadrilateral with exactly one pair of __?__ sides.

5. If the legs of a trapezoid are __?__ , then the trapezoid is an isosceles trapezoid.

6. The parallel sides of a trapezoid are the __?__ .

7. The nonparallel sides of a trapezoid are the __?__ .

**Find the length of the midsegment $\overline{MN}$ of the trapezoid.**

8.
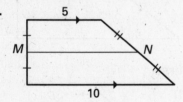

9.
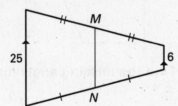

10.
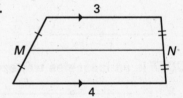

**Find the value of the variable(s).**

11.

12.
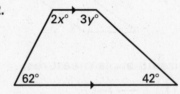

13.
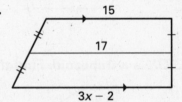

14.
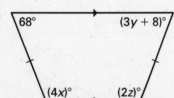

15.
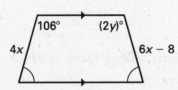

16.
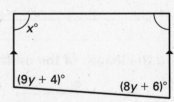

**The vertices of a trapezoid are $A(2, 2)$, $B(5, 2)$, $C(2, 4)$ and $D(5, 6)$.**

17. Plot the vertices in a coordinate plane. Connect them to form trapezoid *CDBA*.

18. Name the bases of trapezoid *CDBA*.

19. State the length of each base.

20. State the length of the midsegment of trapezoid *CDBA*.

# Reteaching with Practice

For use with pages 331–336

**GOAL**  Use properties of trapezoids.

---

### VOCABULARY

A **trapezoid** is a quadrilateral with exactly one pair of parallel sides. The parallel sides are called the **bases**. The nonparallel sides are called the **legs**.

A trapezoid has two pairs of **base angles**.

If the legs of a trapezoid are congruent, then the trapezoid is an **isosceles trapezoid**.

The **midsegment of a trapezoid** is the segment that connects the midpoints of its legs.

**Theorem 6.12**
If a trapezoid is isosceles, then each pair of base angles is congruent.

**Theorem 6.13**
If a trapezoid has a pair of congruent base angles, then it is isosceles.

---

**EXAMPLE 1**  *Find Angle Measures of Trapezoids*

*ABCD* is a trapezoid.
Find the missing angle measures.

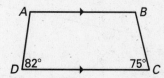

### SOLUTION

By definition, a trapezoid has exactly one pair of parallel sides. In trapezoid *ABCD*, $\overline{AB} \parallel \overline{CD}$. Because $\angle A$ and $\angle D$ are same-side interior angles formed by parallel lines, they are supplementary. So, $m\angle A = 180° - m\angle D = 180° - 82° = 98°$.

Because $\angle B$ and $\angle C$ are same-side interior angles formed by parallel lines, they are supplementary. So, $m\angle B = 180° - m\angle C = 180° - 75° = 105°$.

### Exercises for Example 1

*EFGH* **is a trapezoid. Find the missing angle measures.**

1.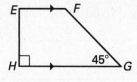

2.

3.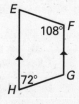

NAME_____ DATE _____

# Reteaching with Practice

For use with pages 331–336

**EXAMPLE 2** **Using Theorem 6.12**

*ABCD* is an isosceles trapezoid. Find the values of *x* and *y*.

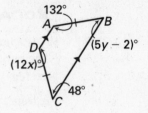

**SOLUTION**

By Theorem 6.12, each pair of base angles in an isosceles trapezoid is congruent. In trapezoid *ABCD*, ∠*A* and ∠*D* are a pair of base angles, and ∠*B* and ∠*C* are a pair of base angles.

| | |
|---|---|
| $m\angle A = m\angle D$ | Theorem 6.12 |
| $132° = 12x°$ | Substitute 132° for $m\angle A$ and 12x° for $m\angle D$. |
| $11 = x$ | Divide each side by 12. |
| $m\angle B = m\angle C$ | Theorem 6.12 |
| $(5y - 2)° = 48°$ | Substitute $(5y - 2)°$ for $m\angle B$ and 48° for $m\angle C$. |
| $5y = 50$ | Add 2 to each side. |
| $y = 10$ | Divide each side by 5. |

## Exercises for Example 2

**Find the values of the variables.**

**4.** isosceles trapezoid *EFGH*

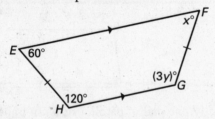

**5.** isosceles trapezoid *JKLM*

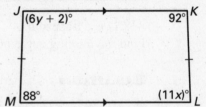

**EXAMPLE 3** **Midsegment of a Trapezoid**

Find the length of the midsegment $\overline{AB}$ of trapezoid *JKLM*.

**SOLUTION**

Use the formula for the midsegment of a trapezoid.

| | |
|---|---|
| $AB = \frac{1}{2}(JK + LM)$ | Formula for midsegment of a trapezoid |
| $AB = \frac{1}{2}(17 + 13)$ | Substitute 17 for *JK* and 13 for *LM*. |
| $AB = 15$ | Simplify. |

## Exercises for Example 3

**Find the length of the midsegment $\overline{AB}$ of the trapezoid.**

**6.**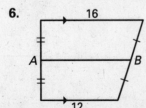

**7.**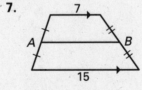

**8.**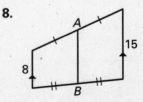

NAME_____ DATE _____

# Quick Catch-Up for Absent Students

**For use with pages 331–336**

The items checked below were covered in class on (date missed) _____

**Technology Activity 6.5: Midsegment of a Trapezoid (p. 331)**

____ **Goal:** Find some properties of the midsegment of a trapezoid.

____ Student Help: Vocabulary Tip

**Lesson 6.5: Trapezoids (pp. 332–333)**

____ **Goal:** Use properties of trapezoids.

*Material Covered:*

____ Example 1: Find Angle Measures of Trapezoids

____ Student Help: Vocabulary Tip

____ Example 2: Midsegment of a Trapezoid

*Vocabulary:*

trapezoid, p. 332                          base angles of a trapezoid, p. 332

bases of a trapezoid, p. 332          isosceles trapezoid, p. 332

legs of a trapezoid, p. 332            midsegment of a trapezoid, p. 333

____ Other (specify) _____

_____

**Homework and Additional Learning Support**

____ Textbook exercises (teacher to specify) pp. 334–336 _____

_____

____ Internet: Homework Help at classzone.com

____ *Reteaching with Practice* worksheet

TEACHER'S NAME _____  CLASS _____  ROOM _____  DATE _____

## *Lesson Plan*

2-day lesson (See *Pacing the Chapter*, TE page 300A)                    **For use with pages 337–341**

**GOAL**  **Identify special quadrilaterals based on limited information.**

State/Local Objectives _____

_____

## ✓ Check the items you wish to use for this lesson.

### STARTING OPTIONS
____ Homework Check (6.5): TE page 334; Answer Transparencies
____ Homework Quiz (6.5): TE page 336, CRB page 58, or Transparencies
____ Warm-Up: CRB page 58 or Transparencies

### TEACHING OPTIONS
____ Examples: Day 1: 1, SE page 337; Day 2: 2–3, SE page 338
____ Extra Examples: TE page 338; Internet Help at classzone.com
____ Checkpoint Exercises: Day 2: 1–3, SE page 338
____ Technology Keystrokes for Ex. 24 on SE page 340: CRB page 59
____ Concept Check: TE page 338
____ Guided Practice Exercises: Day 1: 1–4, SE page 339

### APPLY/HOMEWORK
**Homework Assignment**
____ Basic: Day 1: pp. 339–341 Exs. 5–11, 20–23, 26–38 even
     Day 2: pp. 339–341 Exs. 12–17, 27–37 odd, Quiz 2
____ Average: Day 1: pp. 339–341 Exs. 5–11, 18–24, 26, 35–38
     Day 2: pp. 339–341 Exs. 12–17, 27–34, Quiz 2
____ Advanced: Day 1: pp. 339–341 Exs. 5–11, 18–24, 25*, 26, 35–38
     Day 2: pp. 339–341 Exs. 12–17, 27–34, Quiz 2; EC: classzone.com

### Reteaching the Lesson
____ Practice Masters: CRB pages 60–61 (Level A, Level B)
____ Reteaching with Practice: CRB pages 62–63 or Practice Workbook with Examples;
     Resources in Spanish

### Extending the Lesson
____ Real-Life Application: CRB page 65
____ Challenge: SE page 341; classzone.com

### ASSESSMENT OPTIONS
____ Daily Quiz (6.6): TE page 341 or Transparencies
____ Standardized Test Practice: SE page 341; Transparencies
____ Quiz (6.4–6.6): SE page 341; CRB page 66

Notes _____

_____

_____

TEACHER'S NAME _____ CLASS _____ ROOM _____ DATE _____

# Lesson Plan for Block Scheduling
1-block lesson (See *Pacing the Chapter,* TE page 300A)      For use with pages 337–341

**GOAL**   Identify special quadrilaterals based on limited information.

State/Local Objectives _____

_____

_____

✓ **Check the items you wish to use for this lesson.**

## STARTING OPTIONS
____ Homework Check (6.5): TE page 334; Answer Transparencies
____ Homework Quiz (6.5): TE page 336, CRB page 58,
      or Transparencies
____ Warm-Up: CRB page 58 or Transparencies

## TEACHING OPTIONS
____ Examples: 1–3, SE pages 337–338
____ Extra Examples: TE page 338; Internet Help at classzone.com
____ Checkpoint Exercises: 1–3, SE page 338
____ Technology Keystrokes for Ex. 24 on SE page 340: CRB page 59
____ Concept Check: TE page 338
____ Guided Practice Exercises: 1–4, SE page 339

## APPLY/HOMEWORK
**Homework Assignment**
____ Block Schedule: pp. 339–341 Exs. 5–24, 26–38, Quiz 2

**Reteaching the Lesson**
____ Practice Masters: CRB pages 60–61 (Level A, Level B)
____ Reteaching with Practice: CRB pages 62–63 or Practice Workbook with Examples;
      Resources in Spanish

**Extending the Lesson**
____ Real-Life Application: CRB page 65
____ Challenge: SE page 341; classzone.com

## ASSESSMENT OPTIONS
____ Daily Quiz (6.6): TE page 341 or Transparencies
____ Standardized Test Practice: SE page 341; Transparencies
____ Quiz (6.4–6.6): SE page 341; CRB page 66

Notes _____

_____

_____

| CHAPTER PACING GUIDE | |
|---|---|
| **Day** | **Lesson** |
| 1 | 6.1 |
| 2 | 6.2 |
| 3 | 6.3 |
| 4 | 6.4 |
| 5 | 6.5 |
| 6 | **6.6** |
| 7 | Ch. 6 Review and Assess |

NAME _____ DATE _____

# WARM-UP EXERCISES

For use before Lesson 6.6, pages 337–341

## Name the figure described.

1. a quadrilateral that is both a rhombus and a rectangle

2. a quadrilateral with both pairs of opposite sides parallel

3. a quadrilateral with exactly one pair of parallel sides

4. any parallelogram with perpendicular diagonals

......................................................................................

# DAILY HOMEWORK QUIZ

For use after Lesson 6.5, pages 331–336

## *ABCD* is an isosceles trapezoid.

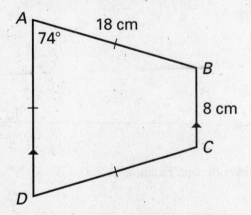

1. Find the length of $\overline{CD}$.

2. Find $m\angle D$.

3. Find $m\angle B$.

4. Name the legs of the trapezoid.

5. Find the length of the midsegment of the trapezoid.

LESSON
# 6.6

NAME _____ DATE _____

## Technology Keystrokes

**For use with Exercise 24, page 340**

Lesson 6.6

### Keystrokes for Exercise 24

### TI-92

**24.** Draw triangle $ABC$.

> **F2** 5 (Move cursor to desired location.) **ENTER** $A$

> (Move cursor to desired location.) **ENTER** $B$ **ENTER**

> (Move cursor to desired location.) **ENTER** $C$ **ENTER**

> (Move cursor to point $A$.) **ENTER**

Draw the midpoint of each side of triangle $ABC$.

> **F4** 3 (Move cursor to $\overline{AB}$.) **ENTER** $D$ (Move cursor to $\overline{BC}$.) **ENTER** $E$

> (Move cursor to $\overline{CA}$.) **ENTER** $F$

Construct $\overline{DF}$ and $\overline{EF}$. **F2** 5 (Move cursor to point $D$.) **ENTER**

> (Move cursor to point $F$.) **ENTER** **ENTER** (Move cursor to point $E$.) **ENTER**

### SKETCHPAD

**24.** Draw triangle $ABC$. Choose the segment tool. Draw $\overline{AB}$, $\overline{BC}$, and $\overline{CA}$ to form triangle $ABC$.

Draw the midpoints of each side of triangle $ABC$. Choose the selection arrow tool. Select $\overline{CA}$. Hold down the shift key and select $\overline{BC}$ and $\overline{AB}$. Choose **Point At Midpoint** from the **Construct** menu.

Draw $\overline{DF}$ and $\overline{EF}$. Choose the segment tool. Draw $\overline{DF}$ and $\overline{EF}$.

NAME _____ DATE _____

## *Practice A*
For use with pages 337–341

**Use the information given in the figures at the right to determine if the following statements are *always* true. Explain your reasoning.**

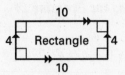
10
4 Rectangle 4
10

1. In a rectangle, all sides are congruent.

2. In a rhombus, all sides are congruent.

Square 5
5    5
5

3. In a trapezoid, the legs are congruent.

4. In a parallelogram, opposite sides are congruent.

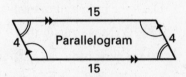
15
4 Parallelogram 4
15

5. In a parallelogram, all angles are congruent.

6. In a rhombus, all angles are congruent.

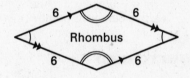
6      6
Rhombus
6      6

7. In a trapezoid, each pair of base angles is congruent.

8. In a square, all angles are congruent.

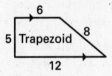
6
5 Trapezoid 8
12

**Determine whether the quadrilateral *ABCD* is a *trapezoid, parallelogram, rectangle, rhombus,* or *square*.**

9.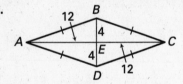

10.

11.

12.

13.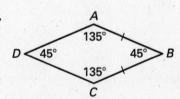

14.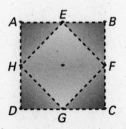

**The figure at the right shows a quilt design.**

15. If *ABCD* is a parallelogram with $AB = BC = CD = DA$, can you conclude that *ABCD* is a square? Explain your reasoning.

16. If *EFGH* is a rectangle with $EF = FG = GH = HE$, can you conclude that *EFGH* is a square? Explain your reasoning.

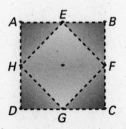

Lesson 6.6

NAME _____ DATE _____

## *Practice B*

For use with pages 337–341

**Put a ✓ mark in the box if the shape always has the given property.**

| | Property | ▱ | Rectangle | Rhombus | Square | Trapezoid |
|---|---|---|---|---|---|---|
| **1.** | Both pairs of opposite sides are congruent. | ? | ? | ? | ? | ? |
| **2.** | Both pairs of opposite sides are parallel. | ? | ? | ? | ? | ? |
| **3.** | All sides are congruent. | ? | ? | ? | ? | ? |
| **4.** | Exactly one pair of opposite sides are parallel. | ? | ? | ? | ? | ? |
| **5.** | Both pairs of opposite angles are congruent. | ? | ? | ? | ? | ? |

**6.** Name all quadrilaterals in which diagonals are congruent.

**7.** Name all quadrilaterals in which diagonals are perpendicular.

**8.** Name all quadrilaterals in which diagonals bisect each other.

**Determine whether the quadrilateral is a *trapezoid, parallelogram, rectangle, rhombus,* or *square*.**

**9.**
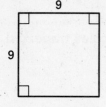

**10.**
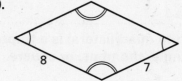

**11.**
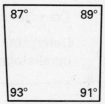

**12.**
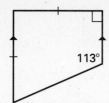

**13.**
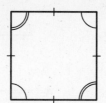

**14.**
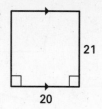

**Are you given enough information to conclude that the figure is the given type of special quadrilateral? Explain your reasoning.**

**15.** A rhombus?
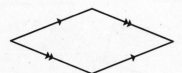

**16.** A rectangle?
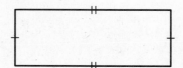

**17.** A trapezoid?
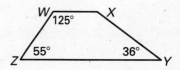

NAME_____ DATE _____

# *Reteaching with Practice*

For use with pages 337–341

**GOAL** Identify special quadrilaterals based on limited information.

**EXAMPLE 1** *Use Properties of Quadrilaterals*

Determine whether the quadrilateral is a trapezoid, isosceles trapezoid, parallelogram, rectangle, rhombus, or square.

a.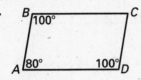

b.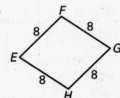

## SOLUTION

**a.** The diagram shows that $\angle A$ is supplementary to $\angle B$ and to $\angle D$. Since one angle of the quadrilateral is supplementary to both of its consecutive angles, you know that $ABCD$ is a parallelogram by Theorem 6.8.

**b.** The diagram shows that all four sides of quadrilateral $EFGH$ have length 8. Since all four sides are congruent, $EFGH$ is a rhombus.

*Exercises for Example 1*

**Determine whether the quadrilateral is a trapezoid, isosceles trapezoid, parallelogram, rectangle, rhombus, or square.**

1.

2.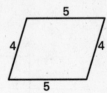

3.

4.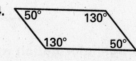

5.

6.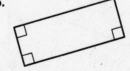

NAME _____ DATE _____

# *Reteaching with Practice*

**For use with pages 337–341**

**EXAMPLE 2**  *Identify a Quadrilateral*

Are you given enough information to conclude that the figure is the given type of special quadrilateral? Explain your reasoning.

**a.** A rectangle?  **b.** A square?  **c.** A parallelogram?

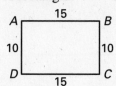 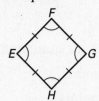 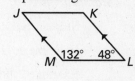

## SOLUTION

**a.** The diagram shows that both pairs of opposite sides are congruent. Therefore, you know that *ABCD* is a parallelogram. For *ABCD* to be a rectangle, all four angles must be right angles. The diagram does not give any information about the angle measures, so you cannot conclude that *ABCD* is a rectangle.

**b.** The diagram shows that all four sides are congruent. Therefore, you know that *EFGH* is a rhombus. For *EFGH* to be a square, all four sides must be congruent and all four angles must be right angles. By the Quadrilateral Interior Angles Theorem, you know that the sum of the measures of the four angles must equal 360°. From the diagram, you know that all four angles have the same measure.

$$m\angle E + m\angle F + m\angle G + m\angle H = 360° \quad \text{Quadrilateral Interior Angles Theorem}$$
$$x° + x° + x° + x° = 360° \quad \text{Substitute } x° \text{ for each angle measure.}$$
$$4x = 360 \quad \text{Simplify.}$$
$$x = 90 \quad \text{Divide each side by 4.}$$

Because all four angles are right angles and all four sides are congruent, you know that *EFGH* is a square.

**c.** The diagram shows that one pair of opposite sides is parallel and the one pair of consecutive angles is supplementary (132° + 48° = 180°). There is no information given about the second pair of opposite sides, nor is there any information given about any other pair of consecutive angles. Therefore, you cannot conclude that *JKLM* is a parallelogram.

### *Exercises for Example 2*

**Are you given enough information to conclude that the figure is the given type of special quadrilateral? Explain your reasoning.**

**7.** An isosceles trapezoid?  **8.** A rhombus?  **9.** A parallelogram?

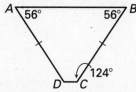 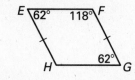 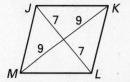

NAME_____ DATE _____

# Quick Catch-Up for Absent Students

**For use with pages 337–341**

The items checked below were covered in class on (date missed) _____

**Lesson 6.6: Reasoning About Special Quadrilaterals (pp. 337–338)**

____ **Goal:** Identify special quadrilaterals based on limited information.

*Material Covered:*

____ Student Help: Study Tip

____ Example 1: Use Properties of Quadrilaterals

____ Example 2: Identify a Rhombus

____ Example 3: Identify a Trapezoid

____ Other (specify) _____

_____

**Homework and Additional Learning Support**

____ Textbook exercises (teacher to specify) <u>pp. 339–341</u>_____

_____

____ Internet: More Examples at classzone.com

____ *Reteaching with Practice* worksheet

NAME_____ DATE _____

# Real-Life Application: When Will I Ever Use This?

For use with pages 337–341

## United States of America

After the Revolutionary War, the United States government had to decide how to organize the settlement of the land they acquired from England. Known as the Northwest Territory, this land was north of the Ohio River and east of the Mississippi River. In 1787 Congress passed the Northwest Ordinance, which established the standard by which much westward expansion took place. The ordinance described a plan by which territories could become states. When the population of a territory reached 60,000 and a state constitution was written and approved by Congress, statehood could be obtained.

The original Northwest Territory became the states of Ohio, Indiana, Illinois, Michigan, and Wisconsin, but a total of thirty-one states originated under the provisions of the Northwest Ordinance over several decades. Expansion was virtually continuous until 1912 when the last contiguous state, Arizona, was added. The last two of the current fifty states, Alaska and Hawaii, were not added until 1959.

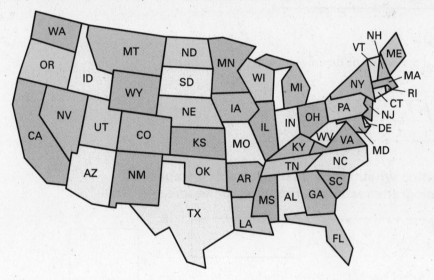

**In Exercises 1–5, use the figure above. Which states (if any) appear to have the given shape?**

1. Parallelogram       2. Rhombus       3. Rectangle
4. Square              5. Trapezoid

NAME _____ DATE _____

## Quiz 2

**For use after Lessons 6.4–6.6**

**1.** *ABCD* is a rhombus. Find the value of *y*.

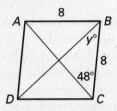

**2.** *EFHG* is a rectangle and *IF* = 8. Find *EH*.

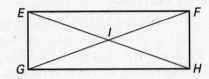

**Answers**

1. _____
2. _____
3. _____
4. _____
5. _____
6. _____
7. _____
8. _____

**In Exercises 3 and 4, use the figure below.**

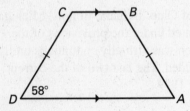

**3.** Find $m\angle A$.

**4.** Find $m\angle B$.

**5.** Find the length of the midsegment of the trapezoid.

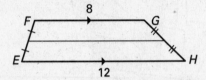

**Determine whether the quadrilateral is a trapezoid, parallelogram, rectangle, rhombus, or square.**

**6.**

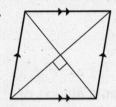

**7.**

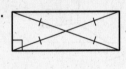

**8.**

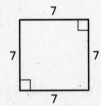

# Chapter Review Games and Activities

**For use after Chapter 6**

**Complete the following number crossword puzzle. All answers are positive integers.**

## Down

1. In a parallelogram, the measure of an angle that is consecutive to an angle that has a measure of 6°.

3. The length of the midsegment of a trapezoid with bases of length 3664 and 2500.

5. The sum of the measures of the interior angles of a quadrilateral.

7. 83 times the number of sides in a heptagon.

9. In a parallelogram, the length of a side that is opposite to a side of length 5621.

## Across

2. In a rhombus, the length of the side adjacent to a side with length 736.

4. 100 times the number of sides in a parallelogram.

6. 263 times the number of right angles in a rectangle.

8. In a parallelogram, the measure of an angle consecutive to an angle of measure 75°.

10. The sum of the measures of the interior angles of a triangle.

# Chapter Test A

**For use after Chapter 6**

**Is the figure a polygon? Explain your reasoning.**

1.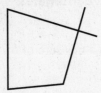

2.

**Find the measure of ∠ A.**

3.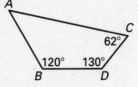

4.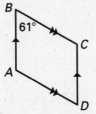

**Complete the statement.**

5. $\overline{AB} \cong$ ___?___

6. $\overline{AD} \cong$ ___?___

7. $\overline{CE} \cong$ ___?___

8. $\overline{EB} \cong$ ___?___

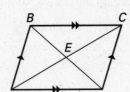

**ABCD is a parallelogram. Find the missing angle measures.**

9.

10.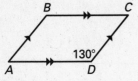

**Decide whether you are given enough information to show the quadrilateral is a parallelogram. Explain your reasoning.**

11.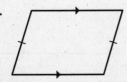

12.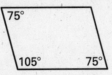

13.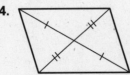

14.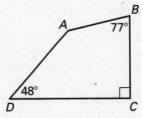

**Answers**

1. _____

_____

2. _____

_____

3. _____

4. _____

5. _____

6. _____

7. _____

8. _____

9. _____

10. _____

11. _____

_____

12. _____

_____

13. _____

_____

14. _____

_____

*Review and Assess*

NAME _____ DATE _____

# Chapter Test A

**For use after Chapter 6**

**Find the measures in rectangle *ABCD*.**

15. Find $m\angle BAD$.

16. Find $AB$.

17. Find $BD$.

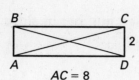
$AC = 8$

**Find the measures in rhombus *EFGH*.**

18. Find $m\angle F$.

19. Find $m\angle G$.

20. Find $EF$.

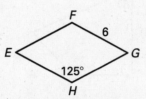

**PQRS is a trapezoid. Find the missing angle measures.**

21.
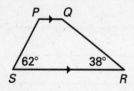

22.
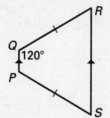

**Find the value of *x* in trapezoid *ABCD*.**

23.
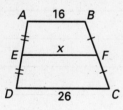

24.
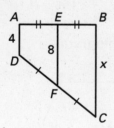

**Are you given enough information to conclude that the figure is the given type of special quadrilateral?**

25. A square?

26. A rectangle?

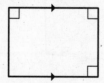

27. An isosceles trapezoid?

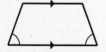

28. A rhombus?

**Answers**

15. _____

16. _____

17. _____

18. _____

19. _____

20. _____

21. _____

22. _____

23. _____

24. _____

25. _____

26. _____

27. _____

28. _____

NAME _____ DATE _____

# Chapter Test B

**For use after Chapter 6**

1. Can a polygon have 2 sides? If so, tell what type of polygon it is.

**Tell what type of polygon has the given number of sides.**

2. 7 sides

3. 13 sides

**Find the value of x.**

4.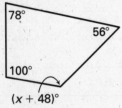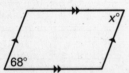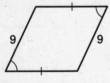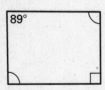

5.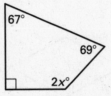

6.

7.

**Find the given measures in □EFGH.**

8. Find EH.

9. Find EF.

10. Find DE.

11. Find FH.

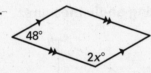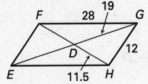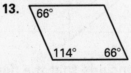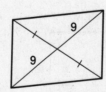

**Tell whether the quadrilateral is a parallelogram. Explain your reasoning.**

12.

13. 66°
114°    66°

14. 89°

15.

**Answers**

1. _____

2. _____

3. _____

4. _____

5. _____

6. _____

7. _____

8. _____

9. _____

10. _____

11. _____

12. _____

13. _____

14. _____

15. _____

_____

NAME _____   DATE _____

# Chapter Test B

**For use after Chapter 6**

**Use quadrilateral *ABCD* shown at the right.**

16. If *ABCD* is a rectangle and *BD* = 6, find *AC*.

17. If *m∠CED* = 92°, is *ABCD* a rhombus?

18. If *ABCD* is a square and *AE* = 4, find *BD*.

19. If *ABCD* is a square, find *m∠BCD*.

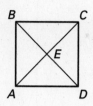

**Find the value of the variable.**

20.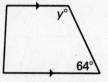

21.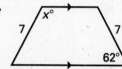

22.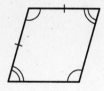

23.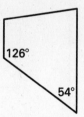

**Are you given enough information to conclude that the figure is the given type of special quadrilateral? Explain your reasoning.**

24. A rhombus?

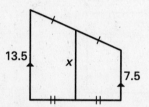

25. A trapezoid?

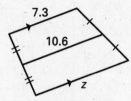

26. A square?

27. A rectangle?

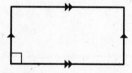

**Answers**

16. _____

17. _____

18. _____

19. _____

20. _____

21. _____

22. _____

23. _____

24. _____

_____

25. _____

_____

26. _____

_____

27. _____

_____

Review and Assess

# SAT/ACT Chapter Test

**For use after Chapter 6**

**1.** Find the measure of $\angle A$.

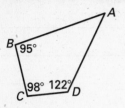

    Ⓐ  45°        Ⓑ  55°
    Ⓒ  58°        Ⓓ  85°

**2.** *ABCD* is a parallelogram. Find $m\angle C$.

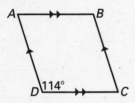

    Ⓐ  57°        Ⓑ  66°
    Ⓒ  76°        Ⓓ  114°

**3.** *KLMN* is a parallelogram. Which of the following statements is false?

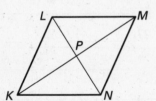

    Ⓐ  *KP = MP*     Ⓑ  *LP = NP*
    Ⓒ  *LM = KN*     Ⓓ  *LN = KM*

**4.** If $\angle D \cong \angle F$ and $\angle E$ and $\angle F$ are supplementary, then you know that *DEFG* must be what type of special quadrilateral?

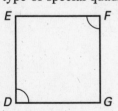

    Ⓐ  parallelogram    Ⓑ  square
    Ⓒ  rhombus         Ⓓ  rectangle

**5.** Which quadrilateral is a parallelogram?

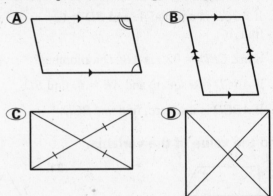

**6.** Which of the following terms could *not* be used to describe the figure below?

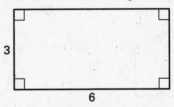

    Ⓐ  parallelogram    Ⓑ  square
    Ⓒ  rectangle        Ⓓ  quadrilateral

**7.** What is the length of the midsegment of trapezoid *ABCD*?

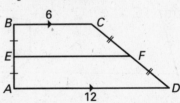

    Ⓐ  6          Ⓑ  8
    Ⓒ  9          Ⓓ  15

**8.** Which of the following statements is *never* true?

    Ⓐ  A trapezoid is a rectangle.
    Ⓑ  A rectangle is a parallelogram.
    Ⓒ  A parallelogram is a rhombus.
    Ⓓ  A square is a rhombus.

*Review and Assess*

**JOURNAL**  **1.** In this chapter, you learned five ways to show that a quadrilateral is
a parallelogram. Describe each method. Be sure to provide a diagram
with appropriate markings for each method.

**MULTI-STEP**  **2.** In the diagram below, *EFGH* is a parallelogram, *BCEH* is a
**PROBLEM**    rectangle, *ADEH* is an isosceles trapezoid, and $AD = 2 \cdot BC$.

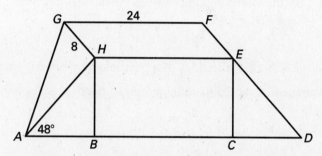

    **a.** Find the measure of ∠*HBC*. Explain your reasoning.

    **b.** Use a theorem about isosceles trapezoids to find the measure of
        ∠*FDC*. What theorem did you use?

    **c.** Explain why the measure of ∠*HEF* is 48°.

    **d.** Use a property of parallelograms to find the measure of ∠*EFG*.

    **e.** Find *HE*. What theorem did you use?

    **f.** Find *BC*. Explain your reasoning.

    **g.** Find *AD*. Explain your reasoning.

    **h.** *Critical Thinking*  Add the midpoint of $\overline{AH}$ to the diagram and
        label it *X*. Add the midpoint of $\overline{DE}$ and label it *Y*. Find the length
        of $\overline{XY}$.

    **i.** *Writing*  What type of triangle is △*AGH*? Describe in words the
        steps you took to find the answer.

**JOURNAL SOLUTION**

**1.** Complete answers should include:

- A clear and concise description of the five ways to show that a quadrilateral is a parallelogram.

- Five clearly labeled diagrams.

**MULTI-STEP PROBLEM SOLUTION**

**2. a.** 90°; the angles of a rectangle are right angles.

**b.** 48°; Theorem 6.12: Base angles of an isosceles trapezoid are congruent.

**c.** Because *BCEH* is a rectangle, and therefore a parallelogram, $\overline{HE} \parallel \overline{BC}$. By the Corresponding Angles Postulate, $m\angle HEF = m\angle FDC = 48°$.

**d.** 132°

**e.** 24; Theorem 6.2: Opposite sides of a parallelogram are congruent.

**f.** 24; a rectangle is a type of parallelogram, so opposite sides are congruent.

**g.** 48; $AD = 2 \cdot BC = 2 \cdot 24 = 48$

**h.** 36

**i.** obtuse; $m\angle AHG + m\angle GHE + m\angle EHB + m\angle BHA = 360°$, so $m\angle AHG + 132° + 90° + 42° = 360°$, and $m\angle AHG = 96°$. Because $\triangle AGH$ contains one obtuse angle, it is an obtuse triangle.

**MULTI-STEP PROBLEM RUBRIC**

**4** Students answer all parts of the problem correctly and completely, showing all work. Students show that they understand the theorems about special quadrilaterals.

**3** Students answer all parts of the problem. Students show that they know the theorems about special quadrilaterals, but may have made minor mathematical errors.

**2** Students answer all parts of the problem. Students may have several mathematical errors. Explanations may be incorrect. Students know some of the theorems about special quadrilaterals.

**1** Students do not complete the problem. Solutions are incorrect. Students do not know the theorems about special quadrilaterals.

*Review and Assess*

NAME_____ DATE _____

# Project: "Puzzling" Shapes

**For use after Chapter 6**

**OBJECTIVE** Use a tangram puzzle set to create polygons and analyze their measurements.

**MATERIALS** ruler, thin cardboard or construction paper, scissors, and a small plastic bag; or a set of tangrams

**INVESTIGATION** Tangrams are considered to be an ancient Chinese puzzle. A tangram puzzle set can be made by cutting a square into the pieces indicated in the diagram.

*Exploring Tangrams*

1. What do you notice about each of the following?

   **a.** the two large right triangles
   **b.** the legs of the medium right triangle
   **c.** the longer sides of the parallelogram
   **d.** the sides of the square and the legs of the small right triangles

2. Create your own tangram puzzle set. Draw a 4-inch by 4-inch square and the remaining line segments as shown in the diagram above.

3. What are the polygons with the fewest sides and with the most sides that you can build using all seven pieces? Name your polygons based on the number of sides they have. Draw a sketch to support your answers.

4. Is it possible to form each of the shapes listed below using exactly two tangram pieces? three pieces? four pieces? five pieces? six pieces? seven pieces? Draw a small sketch of each one you formed, showing the pieces you used.

   **a.** square                              **b.** rectangle that is not a square
   **c.** parallelogram that is not a rectangle   **d.** trapezoid

**PRESENT YOUR RESULTS** Your project report should include all of your sketches and measurement data. Discuss how you approached the problems and how successful you were in finding an approach that helped you find as many polygons as possible. What techniques did you use to organize your work?

Review and Assess

NAME_____ DATE _____

# *Project: Teacher's Notes*

For use after Chapter 6

**GOALS**
- Identify, name, and describe polygons.
- Find the areas of polygons.

**MANAGING THE PROJECT**

Students enjoy working in pairs when using the tangram pieces. This is also a good project for students to share with family. Students may or may not be familiar with the tangram puzzle pieces. Once the diagram for the puzzle has been analyzed and the students have their own set, have them explore the relationships of the pieces to each other and to the original tangram square. Also, you might want to have students create some shapes to get them started. Encourage students to look for non-obvious solutions, such as positioning pieces so that they only touch at their vertices. You might also want to review how to name polygons with many sides.

**RUBRIC**

**The following rubric can be used to assess student work.**

**4** The student identified the triangle as the shape that could be created with the fewest sides and also built and named a polygon with at least 12 sides. The student drew sketches to show how each type of quadrilateral in Exercise 4 could be created using from two to seven tangram pieces. The report indicated that the student developed an organized, thoughtful strategy for finding polygons.

**3** The student identified the triangle as the shape that could be created with the fewest sides and also built and named a polygon with at least ten sides. The student drew sketches to show how the types of quadrilaterals in Exercise 4 could be created using from two to seven tangram pieces, but was missing a few of the answers. The report indicated that the student developed a somewhat organized strategy for finding polygons.

**2** The student was unable to find the polygon with the fewest sides (the triangle) or did not find a polygon with ten or more sides. Students drew sketches to show how some of the quadrilaterals in Exercise 4 could be created using from two to seven tangram pieces. Student summary did not show any evidence of an organized strategy.

**1** Students created and named one shape for Exercise 3. Students drew sketches to show how only a few of the quadrilaterals in Exercise 4 could be created using from two to seven tangram pieces. Summary is missing.

NAME_____ DATE _____

## Cumulative Review

**For use after Chapters 1–6**

**1.** Find *AC*. (*Lesson 1.5*)

**2.** Find *DE*. (*Lesson 1.5*)

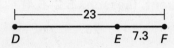

### Use the diagram at the right. (Lessons 1.3, 1.4, and 3.1)

**3.** Name three collinear points.

**4.** Name three noncollinear points.

**5.** Name a pair of skew lines.

**6.** Name the intersection of $\overleftrightarrow{EF}$ and plane *M*.

**7.** Name the intersection of $\overleftrightarrow{AB}$ and plane *M*.

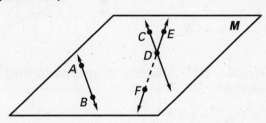

### *M* is the midpoint of $\overline{AB}$. Find the value of *x*. (Lesson 2.1)

**8.**

**9.**

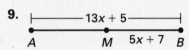

### Use the diagram at the right to find the angle measures. (Lessons 2.4, 3.2, and 3.4)

**10.** $m\angle 1$      **11.** $m\angle 2$

**12.** $m\angle 3$      **13.** $m\angle 4$

**14.** $m\angle 5$      **15.** $m\angle 6$

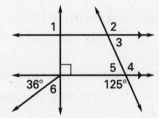

### Find the values of *x* and *y*. Round your answers to the nearest tenth, if necessary. (Lessons 4.3 and 4.4)

**16.**

17  $y°$  $x + 5$
28°

**17.**

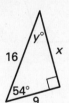

**18.**

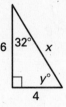

### Determine if the lengths can form a triangle. If so, classify the triangle as *acute*, *obtuse*, or *right*. (Lessons 4.5 and 4.7)

**19.** 6, 12, 19          **20.** 5, 9, 13          **21.** 9, 12, 15

*Review and Assess*

NAME_____ DATE _____

# *Cumulative Review*

**For use after Chapters 1–6**

**State the theorem or postulate that can be used to show the triangles are congruent. (Lessons 5.2–5.4)**

**22.**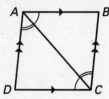

**23.**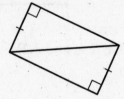

**24.**

$\overrightarrow{TV}$ **bisects** $\angle STU$**. Use the diagrams to find the missing measures. (Lesson 5.6)**

**25.** Find $m\angle STV$ and $SV$.

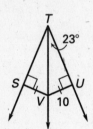

**26.** Find $TU$ and $SV$.

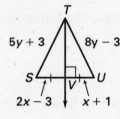

**Name the type of polygon. (Lesson 6.1)**

**27.** 3 sides

**28.** 5 sides

**29.** 7 sides

**Use the diagram to find the measure. (Lessons 6.2–6.5)**

**30.** Find $EB$.

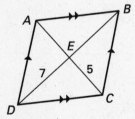

**31.** Find the value of $x$.

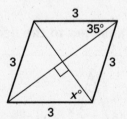

**32.** Find the value of $x$.

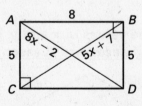

**33.** Find $BE$.

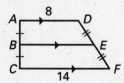

**34.** Find $RU$.

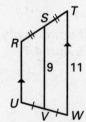

**35.** Find $TS$.

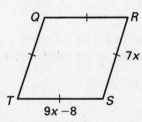

NAME _____ DATE _____

# Cumulative Test

**For use after Chapters 1–6**

1. Show the conjecture is false by finding a counterexample.
   (*Lessons 1.2, 6.5*)

   *Conjecture*: If a quadrilateral has one set of parallel sides, and one
   set of congruent sides, then it must be a rectangle.

**For the given points in the coordinate plane, decide whether
$\overline{AB}$ and $\overline{CD}$ are congruent. (Lesson 1.5)**

2. $A(5, -1)$, $B(5, 6)$, $C(3, 2)$, $D(-4, 2)$

3. $A(-2, -3)$, $B(3, -3)$, $C(8, 0)$, $D(8, 4)$

**Find the measure of the specified angle.  (Lesson 1.6)**

4. Find $m\angle ABC$.

5. Find $m\angle HIK$.

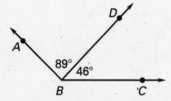

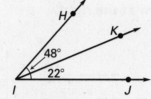

**Find the coordinates of the midpoint of $\overline{KL}$. (Lesson 2.1)**

6. $K(0, -2)$, $L(8, -7)$

7. $K(-5, 1)$, $L(-3, -1)$

8. What is the measure of a compliment to $\angle A$ if
   $m\angle A = 17°$? (*Lesson 2.3*)

**Use the diagram at the right in which $\overrightarrow{HE}$ bisects $\angle FHD$.
(Lessons 2.2–2.4)**

9. Find $m\angle GHF$.

10. Find $m\angle CHD$.

11. Find $m\angle AHB$.

12. Find $m\angle EHD$.

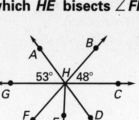

13. Which angle is a supplement of $\angle GHD$?

**Answers**

1. _____

2. _____

3. _____

4. _____

5. _____

6. _____

7. _____

8. _____

9. _____

10. _____

11. _____

12. _____

13. _____

*Review and Assess*

NAME _____ DATE _____

## *Cumulative Test*

**For use after Chapters 1–6**

**In Exercises 14–16, name the property that the statement illustrates. (Lesson 2.6)**

**14.** If the $m\angle A = 52°$, then $m\angle A + 15° = 67°$.

**15.** If $xy = AB$, then $AB = xy$.

**16.** If $m = n$ and $n = T$, then $m = T$.

**17.** Find the value of $x$ in the diagram.
(*Lesson 3.2*)

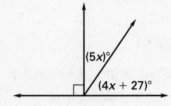

**In Exercises 18–23, use the diagram at the right.**
**(Lessons 3.3, 3.4, 3.6)**

**18.** What kind of angles are $\angle 6$ and $\angle 9$?

**19.** What kind of angles are $\angle 1$ and $\angle 9$?

**20.** What kind of angles are $\angle 4$ and $\angle 14$?

**21.** Find $m\angle 8$.

**22.** Find $m\angle 5$.

**23.** Find $m\angle 6$.

**24.** Using a compass and a straight edge, draw a vertical line $\ell$ and choose a point $P$ to the right of line $\ell$. Construct a line $m$ perpendicular to line $\ell$ through point $P$. (*Lesson 3.6*)

**25.** Find the value of $x$ so that $a \parallel b$. (*Lessons 3.5, 3.6*)

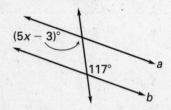

**26.** Describe the translation using coordinate notation. (*Lesson 3.7*)

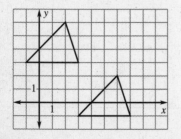

**Answers**

14. _____

_____

15. _____

_____

16. _____

_____

17. _____

18. _____

19. _____

20. _____

21. _____

22. _____

23. _____

24.

25. _____

26. _____

NAME _____ DATE _____

## *Cumulative Test*

For use after Chapters 1–6

**In Exercises 27–29, *C* is the centroid of △*EFG*. (Lesson 4.6)**

27. Find *HF*.

28. Find *EG*.

29. Find *CI*.

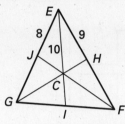

30. Use the Pythagorean theorem
    to find the value of *x*. (*Lesson 4.4*)

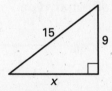

31. Find the distance between *A*(3, −4) and *B*(7, 3).
    Round to the nearest hundredth. (*Lesson 4.4*)

**Find the measure of the numbered angle. Then name the sides of △*ABC* from longest to shortest. (Lessons 4.2, 4.7)**

32.

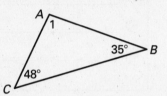

33.

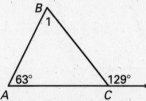

**Decide whether enough information is given to show that the triangles are congruent. If so, state the theorem or postulate that applies. (Lessons 1.5, 5.2–5.5)**

34.

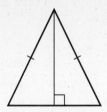

35.

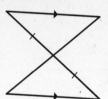

36.

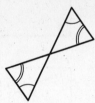

37.

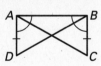

38.

39.

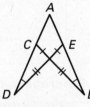

**Use the diagram to find the values of *x* and *y*. (Lesson 5.6)**

40.

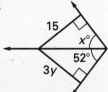

41.

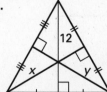

**Answers**

27. _____

28. _____

29. _____

30. _____

31. _____

32. _____

33. _____

34. _____

35. _____

36. _____

37. _____

38. _____

39. _____

40. _____

41. _____

Review and Assess

NAME _____ DATE _____

# *Cumulative Test*

**For use after Chapters 1–6**

**Determine the number of lines of symmetry in the figure. (Lesson 5.7)**

**42.**

**43.**

**Decide if the figure is a polygon. If so, tell what type. If not, explain why. (Lesson 6.1)**

**44.**

**45.**

**46.**

**Find the value of x. (Lessons 6.2–6.4)**

**47.**

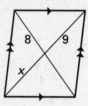

**48.**

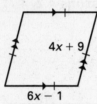

**49.**

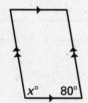

**50.**

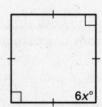

**51.**

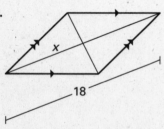

**52.**

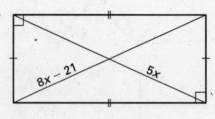

**Find the missing measures in the trapezoid. (Lesson 6.5)**

**53.** Find *FC* and *AF*.

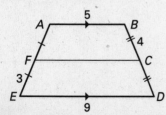

**54.** Find *HM* and *m∠K*.

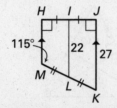

| **Answers** |
| --- |
| 42. _____ |
| 43. _____ |
| 44. _____ |
| 45. _____ |
| 46. _____ |
| 47. _____ |
| 48. _____ |
| 49. _____ |
| 50. _____ |
| 51. _____ |
| 52. _____ |
| 53. _____ |
| 54. _____ |

Review and Assess

# ANSWERS

## Chapter Support

### Parent Guide for Student Success

**6.1:** pentagon   **6.2:** $\overline{AB} \cong \overline{DC}$, $\overline{AD} \cong \overline{BC}$, $\angle ADC$ and $\angle ABC$   **6.3:** *Sample answer*: make sure $AB = DC$ and $AD = BC$; Theorem 6.6 (See p. 319.)   **6.4:** make sure $AC = BD$; Theorem 6.11 (See p. 327.)   **6.5:** 9   **6.6:** always: rectangle, square; sometimes: parallelogram, rhombus

### Strategies for Reading Mathematics

**1.** The symbol $\angle A$ simply refers to the angle with vertex $A$, while the symbol $m\angle A$ means the measure of angle $A$. The symbol $\angle A$ can be used in a congruence statement. The symbol $m\angle A$ can be used to state that the measure of angle $A$ is equal to the measure of another angle or to a number.   **2.** The symbol $\overline{AB}$ simply refers to the segment with endpoints $A$ and $B$, while the symbol $AB$ means the length of $\overline{AB}$. The symbol $\overline{AB}$ can be used in a congruence statement. The symbol $AB$ can be used to state that the length of $\overline{AB}$ is equal to the measure of another segment or to a number.   **3.** The symbol $=$ means "is equal to," while the symbol $\cong$ means "is congruent to." The symbol $=$ can be used to state that the measures of two objects are equal. The symbol $\cong$ can be used to state that two objects are congruent.

**4.** No, the second statement shows different corresponding vertices than the first statement.

## Lesson 6.1

### Warm-Up Exercises

**1.** $180°$   **2.** equilateral triangle   **3.** equiangular triangle

### Daily Homework Quiz

**1.** No; the orientation is not reversed.   **2.** Yes; all three properties of a reflection are met.   **3.** the $y$-axis   **4.** 6

### Practice A

**1.** B   **2.** D   **3.** A   **4.** C   **5.** No; one side is not a segment.   **6.** No; two of the sides intersect only one side.   **7.** Yes; the figure is a polygon formed by six straight sides.   **8.** hexagon

**9.** pentagon   **10.** quadrilateral   **11.** $140°$

**12.** $60°$   **13.** $145°$   **14.** $x = 60$   **15.** $x = 25$

**16.** $x = 30$   **17.** 6 sides; hexagon   **18.** A, B, C, D, E, F   **19.** $\overline{AB}, \overline{BC}, \overline{CD}, \overline{DE}, \overline{EF}, \overline{FA}$

### Practice B

**1.** A, B, C, D, E   **2.** $\overline{AB}, \overline{BC}, \overline{CD}, \overline{DE}, \overline{EA}$

**3.** $\overline{DA}, \overline{DB}$   **4.** The figure is a polygon with four sides, so it is a quadrilateral.   **5.** The figure is not a polygon because two of the sides intersect only one side.   **6.** The figure is not a polygon because some of the sides intersect more than two other sides.   **7.** The figure is not a polygon because one side is not a segment.   **8.** The figure is a polygon with six sides, so it is a hexagon.

**9.** The figure is a polygon with seven sides, so it is a heptagon.   **10.** $77°$   **11.** $61°$   **12.** $70°$

**13.** $x = 20$   **14.** $y = 60$   **15.** $y = 10$

**16.** The figure is an octagon.   **17.** *BCDEFGHA, CDEFGHAB, DEFGHABC, EFGHABCD, FGHABCDE, GHABCDEF, HABCDEFG*

**18.** $\overline{GA}, \overline{GB}, \overline{GC}, \overline{GD}, \overline{GE}$

### Reteaching with Practice

**1.** Yes, the figure is a polygon with eight straight sides.   **2.** Yes, the figure is a polygon with five straight sides.   **3.** No, the figure is not a polygon because it has sides that are not segments.

**4.** Yes, the figure is a polygon with twelve straight sides.   **5.** No, the figure is not a polygon because two of the sides intersect only one other side.   **6.** The figure is a polygon with three sides, so it is a triangle.   **7.** The figure is a polygon with six sides, so it is a hexagon.   **8.** The figure is not a polygon because one of the sides is not a segment.   **9.** $x = 65$   **10.** $x = 9$   **11.** $x = 6$

## Lesson 6.2

### Warm-Up Exercises

1. $16°$   2. $x = 21$

### Daily Homework Quiz

1. No; one side is not a segment.   2. yes; octagon   3. $80°$   4. $108°$   5. $x = 65$

### Practice A

1. yes   2. No; there are five sides.

3. $JK = 6, KL = 4$   4. $JK = 10, KL = 15$

5. $JK = 20, KL = 16$   6. $m\angle W = 75°$, $m\angle X = 105°, m\angle Y = 75°$   7. $m\angle Y = 130°$, $m\angle Z = 50°, m\angle W = 130°$   8. $m\angle X = 80°$, $m\angle Y = 100°, m\angle Z = 80°$   9. 7   10. 11   11. 9

12. $x = 55, y = 125$   13. $x = 8, y = 6$

14. $x = 2.5, y = 4$   15. $\overline{QR}$   16. $\angle PSR$

17. $\overline{SR}$

### Practice B

1. In $\square PQRS$, $\overline{PQ} \cong \overline{SR}$ and $\overline{QR} \cong \overline{PS}$.

2. In $\square PQRS$, $\angle QPS \cong \angle SRQ$ and $\angle PQR \cong \angle RSP$.

3. In $\square PQRS$, $x° + y° = 180°$.

4. In $\square PQRS$, $\overline{QM} \cong \overline{MS}$ and $\overline{PM} \cong \overline{MR}$.

5. $DA = 14, DC = 7$   6. $GE = 6.2, DF = 16.4$

7. $m\angle P = 36°, m\angle Q = 144°, m\angle R = 36°$

8. $GE = 11, GF = 9$

9. $DA = 1.25, DC = 0.75$   10. $m\angle Q = 93°$, $m\angle R = 87°, m\angle S = 93°$   11. $x = 49, y = 131$

12. $x = 2, y = 3$   13. $x = 5, y = 3$

14. $\overline{JK} \cong \overline{MP}, \overline{JM} \cong \overline{KP}$   15. $\angle NMP \cong \angle Q$, $\angle MPQ \cong \angle N$   16. $\angle J$ and $\angle KPM$

### Reteaching with Practice

1. $x = 12, y = 9$   2. $x = 23, y = 6$   3. $x = 1$, $y = 4$   4. $x = 9, y = 135$   5. $x = 55, y = 9$

6. $x = 12, y = 6$   7. $x = 7, y = 15$   8. $x = 1$, $y = 8$   9. $x = 5, y = 3$

## Lesson 6.3

### Warm-Up Exercises

1. Opposite sides of a parallelogram are congruent.   2. The diagonals of a parallelogram bisect each other.   3. Opposite angles of a parallelogram are congruent.   4. Consecutive angles of a parallelogram are supplementary.

### Daily Homework Quiz

1. 8   2. $72°$   3. $108°$   4. 14   5. They are congruent.

### Practice A

1. B   2. D   3. A   4. C   5. Yes; if both pairs of opposite sides of a quadrilateral are $\cong$, then it is a $\square$.   6. No; only one pair of opposite sides are $\cong$.   7. Yes; if both pairs of opposite sides of a quadrilateral are $\cong$, then it is a $\square$.   8. Yes; if both pairs of opposite $\angle$ of a quadrilateral are $\cong$, then it is a $\square$.   9. No; only one pair of opposite $\angle$ is $\cong$.   10. Yes; if both pairs of opposite $\angle$ of a quadrilateral are $\cong$, then it is a $\square$.   11. No; $\angle A$ is supplementary to consecutive $\angle D$ but $\angle A$ is not supplementary to consecutive $\angle B$.

12. Yes; $\angle E$ is supplementary to both consecutive $\angle$, $\angle D$ and $\angle F$.   13. Yes; $\angle M$ is supplementary to both consecutive $\angle$, $\angle N$ and $\angle P$   14. Yes; if the diagonals of a quadrilateral bisect each other, then it is a $\square$.   15. No; only one diagonal is bisected.   16. Yes; if the diagonals of a quadrilateral bisect each other, then it is a $\square$.   17. Show that both pairs of opposite $\angle$ are $\cong$; show that one $\angle$ is supplementary to both of its consecutive $\angle$.

18. Show that both pairs of opposite sides are parallel; show that both pairs of opposite sides are $\cong$.

### Practice B

1. Show that both pairs of opposite sides are parallel; show that both pairs of opposite sides are $\cong$.   2. Show that both pairs of opposite $\angle$ are $\cong$; show that one $\angle$ is supplementary to both of its consecutive $\angle$.   3. Show that the diagonals bisect each other.   4. Yes; if both pairs of opposite $\angle$ of a quadrilateral are $\cong$, then it is a $\square$.

5. No; no pairs of opposite $\angle$ are $\cong$, or no angle is supplementary to both of its consecutive angles.

# Lesson 6.3 *continued*

**6.** Yes; if the diagonals of a quadrilateral bisect each other, then it is a ▱. **7.** Yes; if both pairs of opposites sides of a quadrilateral are ≅, then it is a ▱. **8.** Yes; if an ∠ is supplementary to both of its consecutive ∢, then it is a ▱. **9.** No; the diagonals do not bisect each other. **10.** Yes; if the diagonals bisect each other, then the quadrilateral is a ▱. **11.** No; no pairs of opposite ∢ are ≅, or no angle is supplementary to both of its consecutive angles. **12.** Yes; if opposite sides of a quadrilateral are ≅, then it is a ▱.

**13.** Slope of $\overline{AB}$ = 2, slope of $\overline{DC} = \frac{4}{3}$; quadrilateral *ABCD* is not a ▱. **14.** Slope of $\overline{EF} = \frac{2}{3}$, slope of $\overline{HG} = \frac{2}{3}$; quadrilateral *EFGH* is a ▱. **15.** Yes. At top, 4 right ∢ mean opposite ∢ are ≅, so it is a ▱. Then opposite sides are ≅, which holds true at bottom. So the front edges also form a ▱ in the bottom picture.

## Reteaching with Practice

**1.** The quadrilateral is not a parallelogram. Both pairs of opposite sides are not congruent.

**2.** The quadrilateral is a parallelogram because both pairs of opposite sides are congruent.

**3.** The quadrilateral is not a parallelogram. Both pairs of opposite sides are not congruent.

**4.** The quadrilateral is a parallelogram, because both pairs of opposite angles are congruent.

**5.** The quadrilateral is not a parallelogram. Both pairs of opposite angles are not congruent.

**6.** The quadrilateral is a parallelogram, because both pairs of opposite angles are congruent.

**7.** *ABCD* is a parallelogram, because ∠A is supplementary to ∠B and ∠D.

**8.** The quadrilateral is not a parallelogram. The diagonals do not bisect each other. **9.** *JKLM* is not a parallelogram. ∠K is supplementary to ∠L, but ∠K is not supplementary to ∠J.

## Quiz 1

**1.** hexagon **2.** 126° **3.** 138° **4.** 65° **5.** 115° **6.** 18.5 **7.** 30.8 **8.** Yes; the diagonals of quadrilateral *ABCD* bisect each other, so *ABCD* is a ▱.

**9.** Yes; ∠A is supplementary to both consecutive angles, ∠B and ∠D, so *ABCD* is a ▱.

**10.** No; you don't know ∠A ≅ ∠C, so both pairs of opposite angles are not congruent.

# Lesson 6.4

## Warm-Up Exercises

**1.** $x = 30$  **2.** $x = 3$

## Daily Homework Quiz

**1.** No; consecutive angles are not supplementary.

**2.** Yes; by the definition of parallelogram

**3.** Yes; both pairs of opposite angles are congruent. **4.** Yes; both pairs of opposite sides are congruent.

## Practice A

**1.** rhombus **2.** rectangle **3.** square **4.** ⊥

**5.** ≅ **6. a.** 90° **b.** 90° **c.** 90° **d.** 90°

**7. a.** 6 **b.** 6 **c.** 6

**8. a.** 90° **b.** 5 **c.** 5 **d.** 90° **9.** rhombus

**10.** square **11.** rectangle **12.** $x = 30$

**13.** $x = 2$ **14.** $x = 90$ **15.** rhombus

**16.** rectangle

## Practice B

**1.** If $\overline{AB} \cong \overline{BC} \cong \overline{CD} \cong \overline{AD}$, then *ABCD* is a rhombus. **2.** If $m\angle A = m\angle B = m\angle C = m\angle D = 90°$, then *ABCD* is a rectangle.

**3.** If $\overline{AB} \cong \overline{BC} \cong \overline{CD} \cong \overline{AD}$ and $m\angle A = m\angle B = m\angle C = m\angle D = 90°$, then *ABCD* is a square. **4.** In rhombus *ABCD*, $\overline{AC} \perp \overline{BD}$.

**5.** In rectangle *ABCD*, $\overline{AC} \cong \overline{BD}$. **6. a.** 10.5 **b.** 10.5 **c.** 10.5 **7. a.** 90° **b.** 7 **c.** 2

**8. a.** 90° **b.** 6 **c.** 6 **9.** rectangle, parallelogram, square, rhombus **10.** square, rectangle, parallelogram, rhombus **11.** square **12.** rectangle, square **13.** $x = 40$ **14.** $x = 2$

**15.** $y = 1$ **16.** You can make four 13-inch sides; to be a square, the distance between the opposite corners should be equal.

# Lesson 6.4 *continued*

## Reteaching with Practice

**1.** $x = 7, y = 30$ **2.** $x = 2, y = 4$ **3.** $x = 3$, $y = 13$ **4.** $x = 18$ **5.** $x = 45$ **6.** $x = 8$

**7.** $x = 7$

## Real-Life Application

**1. a.** rhombus **b.** rhombus, rectangle, square
**c.** rectangle

**2.**

| Shape | Rhombus | Rectangle | Square |
|-------|---------|-----------|--------|
| Number | 20 | 30 | 16 |

# Lesson 6.5

## Warm-Up Exercises

**1.** $(2, -1)$ **2.** $(-2, -3)$ **3.** $y = 34$
**4.** $y = 5$

## Daily Homework Quiz

**1.** rhombus **2.** rhombus **3.** rectangle
**4.** rectangle **5.** square

## Practice A

**1.** B **2.** C **3.** E **4.** A **5.** D
**6.** isosceles trapezoid **7.** neither **8.** trapezoid
**9.** $m\angle D = 60°, m\angle E = 120°, m\angle F = 120°$
**10.** $m\angle E = 145°, m\angle F = 145°, m\angle C = 35°$
**11.** $m\angle D = 140°, m\angle E = 40°, m\angle F = 40°$
**12.** $m\angle F = 150°, m\angle H = 55°$
**13.** $m\angle F = 130°, m\angle H = 90°$
**14.** $m\angle E = 75°, m\angle F = 50°$
**15.** 10 **16.** 7 **17.** 8

## Practice B

**1.** congruent **2.** base **3.** midsegment (or median) **4.** parallel **5.** congruent **6.** bases
**7.** legs **8.** 7.5 **9.** 15.5 **10.** 3.5
**11.** $x = 90, y = 103$ **12.** $x = 59, y = 46$
**13.** $x = 7$ **14.** $x = 28, y = 20, z = 56$
**15.** $x = 4, y = 53$ **16.** $x = 90, y = 10$

**17.**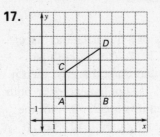

**18.** $\overline{AC}$ and $\overline{BD}$ **19.** $AC = 2, BD = 4$ **20.** 3

## Reteaching with Practice

**1.** $m\angle E = 90°, m\angle F = 135°$ **2.** $m\angle F = 58°$, $m\angle H = 85°$ **3.** $m\angle E = 72°, m\angle G = 108°$
**4.** $x = 60, y = 40$ **5.** $x = 8, y = 15$
**6.** $AB = 14$ **7.** $AB = 11$ **8.** $AB = \frac{23}{2}$

# Lesson 6.6

## Warm-Up Exercises

**1.** square **2.** parallelogram **3.** trapezoid
**4.** rhombus

## Daily Homework Quiz

**1.** 18 cm **2.** 74° **3.** 106° **4.** $\overline{AB}$ and $\overline{DC}$
**5.** 13 cm

## Practice A

**1.** No; in a rectangle, opposite sides are $\cong$. All sides are $\cong$ only if the rectangle is a square.

**2.** yes; true by the definition of a rhombus

**3.** no; only true for isosceles trapezoid

**4.** Yes; opposite sides of a $\square$ are $\cong$.

**5.** No; all angles are $\cong$ only if the $\square$ is a rectangle or square. **6.** no; only true if the rhombus is a square **7.** no; only true for isosceles trapezoid **8.** yes; true by definition of a square **9.** rhombus **10.** square **11.** rectangle
**12.** trapezoid **13.** parallelogram **14.** rhombus
**15.** No; *ABCD* is a rhombus by definition, but you don't know the angles. **16.** Yes; *EFGH* is a rectangle, so it is a parallelogram with four right $\angle$. Because it also has four $\cong$ sides, it is a square by definition.

# Lesson 6.6 *continued*

## Practice B

| | Property | ▱ | Rect. | Rhom. | Sq. | Trap. |
|---|---|---|---|---|---|---|
| 1. | Both pairs of opp. sides are ≅. | ✔ | ✔ | ✔ | ✔ | |
| 2. | Both pairs of opp. sides are ∥. | ✔ | ✔ | ✔ | ✔ | |
| 3. | All sides are ≅. | | | ✔ | ✔ | |
| 4. | Exactly 1 pr. of opp. sides are ∥. | | | | | ✔ |
| 5. | Both pairs of opp. ∠ are ≅. | ✔ | ✔ | ✔ | ✔ | |

**6.** rectangle, square   **7.** rhombus, square

**8.** ▱, rectangle, rhombus, square   **9.** square

**10.** parallelogram   **11.** trapezoid   **12.** trapezoid

**13.** rhombus   **14.** rectangle   **15.** No; because opposite sides are parallel, it is a ▱, but nothing is known about the sides, so you can't say it is a rhombus.   **16.** No; both pairs of opposite sides are ≅, so the figure is a ▱, but no information is given about the measure of the ∠.   **17.** Yes; ∠W is supplementary to ∠Z, therefore $\overline{WX} \parallel \overline{ZY}$ by the Same-Side Interior Angles Converse Theorem. ∠Z and ∠Y are not supplementary, therefore $\overline{ZW}$ is not parallel to $\overline{YX}$. By definition, *WXYZ* is a trapezoid.

## Reteaching with Practice

**1.** isosceles trapezoid   **2.** parallelogram

**3.** trapezoid   **4.** parallelogram   **5.** square

**6.** rectangle   **7.** Yes, *ABCD* is an isosceles trapezoid. Because same-side interior angles ∠B and ∠C are supplementary, you know that opposite sides $\overline{AB}$ and $\overline{CD}$ are parallel. Because same-side interior angles ∠A and ∠B are *not* supplementary, you know that $\overline{AD}$ and $\overline{BC}$ are *not* parallel. Therefore, you know that *ABCD* is a trapezoid since it has exactly one pair of opposite sides parallel. From the diagram you know that the legs $\overline{AD}$ and $\overline{BC}$ are congruent, so you know that *ABCD* is an isosceles trapezoid.

**8.** No, you cannot conclude that *EFGH* is a rhombus. Because ∠F is supplementary to both of its consecutive angles, you know that *EFGH* is a parallelogram. From the diagram you know that two of the sides are congruent, but there is no information given about the other two sides.

**9.** Yes, *JKLM* is a parallelogram. By Theorem 6.9, because the diagonals bisect each other, *JKLM* is a parallelogram.

## Real-Life Application

**1.** WY, ME, SD   **2.** ME   **3.** SD, WY

**4.** none   **5.** WA, ND, TN, IN, VT, CT, KS, RI

## Quiz 2

**1.** $y = 42$   **2.** 16   **3.** 58°   **4.** 122°   **5.** 10

**6.** rhombus   **7.** rectangle   **8.** square

# Review and Assessment

## Chapter Review Games and Activities

### Down

**1.** 174   **3.** 3082
**5.** 360   **7.** 581
**9.** 5621

### Across

**2.** 736   **4.** 400
**6.** 1052   **8.** 105
**10.** 180

# Review and Assessment *continued*

## Test A

1. No; two endpoints don't intersect sides.

2. No; a side is not a segment.   **3.** 48°   **4.** 145°

5. $\overline{DC}$   **6.** $\overline{BC}$   **7.** $\overline{AE}$   **8.** $\overline{ED}$

9. $m\angle A = m\angle C = 50°, m\angle B = 130°$

10. $m\angle A = m\angle C = 119°, m\angle D = 61°$

11. Yes; both pairs of opposite $\angle$ are $\cong$ .

12. Yes; an $\angle$ is supplementary to both consecutive $\angle$ .   **13.** No; only one pair of sides are $\cong$ and one pair of sides are $\parallel$.   **14.** Yes; diagonals bisect each other.

15. 90°   **16.** 2   **17.** 8   **18.** 125°   **19.** 55°

20. 6   **21.** $m\angle P = 118°, m\angle Q = 142°$

22. $m\angle P = 120°, m\angle S = m\angle R = 60°$

23. $x = 21$   **24.** $x = 12$   **25.** no   **26.** yes

27. yes   **28.** yes

## Test B

1. no   **2.** heptagon   **3.** 13-gon   **4.** $x = 78$

5. $x = 67$   **6.** $x = 68$   **7.** $x = 66$   **8.** 28

9. 12   **10.** 19   **11.** 23   **12.** Yes; both pairs of opposite sides are $\cong$.   **13.** Yes; an $\angle$ is supplementary to both its consecutive $\angle$.   **14.** No; opposite $\angle$ are not $\cong$.   **15.** Yes; diagonals bisect each other.   **16.** 6   **17.** no   **18.** 8   **19.** 90°

20. $y = 116$   **21.** $x = 118$   **22.** $x = 10.5$

23. $z = 13.9$   **24.** Yes; it is a $\square$ because opposite $\angle$ are $\cong$ . That means opposite sides are $\cong$ . Because 2 adjacent sides are $\cong$, all 4 sides must be $\cong$.   **25.** Yes; the right and left sides are parallel by Same-Side Interior Angles Converse, and you can see that the top is not $\parallel$ to the bottom.   **26.** Yes; the unmarked $\angle$ measures 90° by Interior Angles of a Quadrilateral. Because opposite $\angle$ are $\cong$ , it is a $\square$. So, opposite sides must be $\cong$, and because 2 adjacent sides are $\cong$ all 4 sides must be $\cong$ .   **27.** Yes; it is a $\square$ by definition, so opposite $\angle$ are $\cong$ and adjacent $\angle$ are supplementary. All 4 $\angle$ are right.

## SAT/ACT Chapter Test

1. A   **2.** B   **3.** D   **4.** A   **5.** B   **6.** B   **7.** C

8. A

## Alternative Assessment

1. Complete answers should include a clear and concise description of the five ways to show that a quadrilateral is a parallelogram and five clearly labeled diagrams.   **2. a.** 90°; the angles of a rectangle are right angles.   **b.** 48°; Theorem 6.12: Base angles of an isosceles trapezoid are congruent.   **c.** Because $BCEH$ is a rectangle, and therefore a parallelogram, $\overline{HE} \parallel \overline{BC}$. By the Corresponding Angles Postulate, $m\angle HEF = m\angle FDC = 48°$.   **d.** 132°

  **e.** 24; Theorem 6.2: Opposite sides of a parallelogram are congruent.   **f.** 24; a rectangle is a type of parallelogram, so opposite sides are congruent.

  **g.** 48; $AD = 2 \cdot BC = 2 \cdot 24 = 48$   **h.** 36

  **i.** obtuse;
$m\angle AHG + m\angle GHE + m\angle EHB + m\angle BHA = 360°$, so $m\angle AHG + 132° + 90° + 42° = 360°$, and $m\angle AHG = 96°$. Because $\triangle AGH$ contains one obtuse angle, it is an obtuse triangle.

## Project: "Puzzling" Shapes

1. **a.** They are congruent.   **b.** They are congruent, with length equal to one-half the length of the side of the entire square.   **c.** Their length is one-half the length of the side of the entire square.   **d.** They are congruent, with length equal to one-fourth the length of the square's diagonal.

3. Triangle; *Sample drawing:*

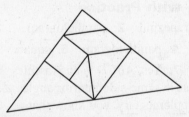

The polygon with the largest number of sides depends on creativity. Check that students have named their polygons correctly.

# Review and Assessment *continued*

**4.** Sample drawings are given for parts (a)–(d). Note that in the sample drawings triangles have been identified as small (S), medium (M), or large (L).

**a.** A square cannot be formed with 6 pieces.

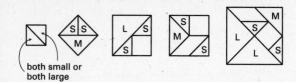

**b.** A rectangle that is not a square cannot be formed with 2 pieces.

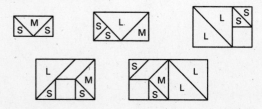

**c.** A parallelogram that is not a rectangle cannot be formed with 6 pieces.

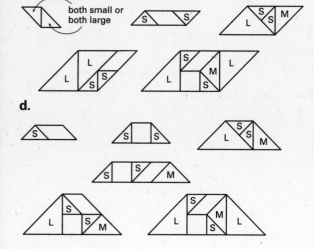

**d.**

## Cumulative Review

**1.** 1  **2.** 15.7  **3.** *E*, *D*, and *F*
**4.** *Sample answer:* A, B, and C  **5.** $\overleftrightarrow{AB}$ and $\overleftrightarrow{DE}$
**6.** D  **7.** $\overleftrightarrow{AB}$  **8.** $x = 7$  **9.** $x = 3$  **10.** 90°
**11.** 125°  **12.** 55°  **13.** 125°  **14.** 55°
**15.** 54°  **16.** $x = 12, y = 124$  **17.** $x \approx 13.2$,
$y = 36$  **18.** $x \approx 7.2, y = 58$  **19.** no  **20.** yes;
obtuse  **21.** yes; right  **22.** ASA  **23.** HL
**24.** AAS  **25.** $m\angle STV = 23°, SV = 10$
**26.** $TU = 13, SV = 5$  **27.** triangle

**28.** pentagon  **29.** heptagon  **30.** 7
**31.** $x = 55$  **32.** $x = 3$  **33.** 11  **34.** 7
**35.** 28

## Cumulative Test

**1.** counterexample: isosceles trapezoid  **2.** yes
**3.** no  **4.** 135°  **5.** 26°  **6.** (4, −4.5)
**7.** (−4, 0)  **8.** 73°  **9.** 48°  **10.** 53°  **11.** 79°
**12.** $39\frac{1}{2}°$  **13.** $\angle AHG$ or $\angle CHD$  **14.** Addition
Property of Equality  **15.** Symmetric Property of
Equality  **16.** Transitive Property of Equality
**17.** $x = 7$  **18.** alternate interior angles
**19.** corresponding angles  **20.** alternate exterior
angles  **21.** 90°  **22.** 65°  **23.** 115°
**24.**

**25.** $x = 24$  **26.** $(x, y) \rightarrow (x + 4, y - 4)$  **27.** 9
**28.** 16  **29.** 5  **30.** $x = 12$  **31.** 8.06
**32.** 97°; $\overline{CB}, \overline{AB}, \overline{AC}$  **33.** 66°; $\overline{AC}, \overline{BC}, \overline{AB}$
**34.** yes; HL  **35.** yes; ASA or AAS  **36.** no
**37.** yes; SAS  **38.** yes; SSS  **39.** yes; AAS
**40.** $x = 52, y = 5$  **41.** $x = 12, y = 12$  **42.** 1
**43.** 2  **44.** yes, octagon  **45.** No, sides intersect
more than two other sides.  **46.** yes, pentagon
**47.** $x = 9$  **48.** $x = 5$  **49.** $x = 100$
**50.** $x = 15$  **51.** $x = 9$  **52.** $x = 7$
**53.** $FC = 7, AF = 3$
**54.** $HM = 17, m\angle K = 65°$

**Answers**